Interpretation
of Metallographic Structures

Interpretation
of Metallographic Structures

THIRD EDITION

WILLIAM ROSTOKER

Rostoker, Inc.
Burham, Illinois

JAMES R. DVORAK

Fond du Lac, Wisconsin

ACADEMIC PRESS, INC.
Harcourt Brace Jovanovich, Publishers
San Diego   New York   Boston
London   Sydney   Tokyo   Toronto

This book is printed on acid-free paper. ∞

Academic Press, Inc.
San Diego, California 92101

United Kingdom Edition published by
Academic Press Limited
24–28 Oval Road, London NW1 7DX

Library of Congress Cataloging-in-Publication Data

Rostoker, William
    Interpretation of metallographic structures / William Rostoker,
James R. Dvorak. -- 3rd ed.
        p.   cm.
    Includes bibliographical references.
    ISBN   0-12-598255-0   (alk. paper)
    1. Metallography.   I. Dvorak, James R. II. Title.
TN690.R75   1990
669'.95--dc20   89-18139
                                                                      CIP

Printed in the United States of America
90   91   92   93   9   8   7   6   5   4   3   2   1

Contents

Preface to Third Edition

Light optical microscopy continues to be the main instrument and technique of metallographic examination. At least, it should be the first reconnoiter. Etched microstructures are now also commonly examined by the scanning electron microscope (SEM). The appearances by either light or electron optical means are much the same. The major differences are in the qualities of contrasts between adjacent phases and at interfaces. In these, light microscopy is more versatile, but the SEM has the advantage of being able to "zoom" to magnifications well in excess of light optical limits.

There are a good many other instruments that have become more widely used to supplement the optical microscope. The "quantometer," or electronic area counting instrument, can now easily provide the volume proportions of a distributed phase. The microhardness tester can indicate differences in state that have not been brought out by etching contrasts. We also add to these the variety of instruments that can provide qualitative and quantitative analysis of the elements present.

It is the microbeam analysis of phases and designated zones in a microstructure that has grown so much in use since the previous edition was published. This book is substantially augmented by the addition of a new chapter that deals with this subject and, in particular, the uses to which it may be put.

As a final word, we wish to dedicate this third edition to the memory of a good friend and colleague, Robert F. Domagala.

Preface to Second Edition

The preparation of this second edition has afforded the opportunity of replacing a number of the specimen microstructures with more appropriate alternatives and, of course, to correct errors in the text that became apparent only after the first edition was published.

A major change in the text has been to treat the subject of fracture in a separate and distinct chapter. The recognition of fracture modes has developed considerably in the past ten years. In particular, the scanning electron microscope has created appearance identities that are often unique and definitive. Accordingly, fracture analysis must use SEM and light optical microscopy jointly. This joint use is the theme of the new chapter.

The subject of quantitative metallography has become very sophisticated, but the application of measurements to property and process correlation has remained rather elementary. Opportunity has been taken in this edition to present some of the types of correlation that have become very useful.

Preface to First Edition

While the modern metallurgist has many investigative techniques at his disposal, the optical reflection microscope remains his most effective means for examining the structure of metals. At the lowest level of use, it provides a ready means for comparison and identification. With skill and experience there is always much more to be gained. The recognition of significant features combined with an appreciation of physical metallurgy provides a powerful basis for rationalization and diagnosis. To teach this art requires carefully selected examples placed within the context of real issues. This book is intended to be such a vehicle. It is directed toward the senior student as a preview of the scope of his subject and to the practicing metallurgist as a reintroduction.

The book is an organized presentation of specimen microstructures, each chosen for its clarity of illustration and each or in groups forming the pretext for some discussion of interrelation between physical metallurgy and metallography. There has been no intent to produce an encyclopedia of microstructures in the sense that all variations of all alloy systems are catalogued. The objective has been, instead, to choose structures characteristic in a physical metallurgy sense with the purpose of demonstrating that logical framework of interpretation can supplant mental storage of infinite variations.

Discussion has been divided into two parts. Prefacing each topic is a brief summary of those aspects of the subject which are relevant to metallographic observation. The selection of micrographs is integrated into this summary. Each micrograph is given an individual commentary in which the composition and thermal or thermal-mechanical history are related to the particular point which is intended for illustration. These short descriptions frequently contain remarks which are apropos of the structure but not necessarily of the topic under discussion. This is deliberately done to encourage the reader to see more than he is looking for.

The final chapter deviates from those preceding by approaching the subject of quantitative metallography. This subject has matured greatly in recent years and quantitative measurements will assuredly become an increasingly important tool in physical metallurgy. Whether or not the reader will have occasion to use these techniques, an understanding of them will help in appreciating the limitations of planar section views as representative of the structure of opaque bodies.

To demonstrate that metallographic interpretation is a common denominator of all metallic materials, specimen microstructures have been deliberately taken from a wide diversity of materials. This may appear to have been carried to an extreme but the authors have felt that great rhetorical value derives from the mixture of common and uncommon.

The microstructures themselves are intended to be technically good and at magnifications consistent with the objectives of the associated text. The quality of presentation has been something of a dilemma. One cannot forestall the reader from congratulating himself on having seen or even prepared specimens esthetically superior to one or another micrograph offered in the text. The authors can only hope that each reader will actually find such occasion. On the other hand it may be said that the presentations are not typical of what the metallurgist most often encounters. This is undoubtedly true and the novice diagnostician using this book as a guide may be disillusioned on occasions of encountering the real-life problems of professional service. At this point, the authors must reiterate their objectives of attempting to teach rather than to catalogue.

The written presentation has attempted to embody contemporary concepts in physical metallurgy without recreating their origin or derivation. For these the reader must turn to the many advanced books on the subject. Occasionally, the text makes footnote references to certain publications which give pertinent elaboration. For more complete details of special techniques, standards, and general information on metallographic technology, a classified list of suitable references is given in the Appendix.

Introduction

The early training of a metallurgist is usually built around metallographic structures. The principles of phase constitution, transformations and thermomechanical histories are most easily described to the student in terms of the appearance of metals under the microscope. Perhaps just because it is so implicit in the training of a metallurgist, metallography is easy to regard as an established art. Yet it can be no more settled or established than the science of metallurgy itself. For practical reasons as well as visual appreciation, each new increment of knowledge should be translated into the nature of the structure of metals.

While metallography remains a versatile tool, modern metallurgy makes use of a wide variety of physical and mechanical property and radiation interactive measurements in the definition of states of aggregation. In sober appraisal there is little room for controversies on the relative merits of various tools. The fact stands out that no tool is self-sufficient, for they each describe different aspects of the nature of a material—aspects whose separate knowledge is inadequate but in joint appraisal may permit unambiguous interpretation. It remains, therefore, to identify properly the limitations as well as the capabilities of the tools of metallurgical study.

Since this book concerns itself with metallography as a metallurgical tool, it behooves the reader to consider the nature of what may be observed with the aid of an optical reflection microscope. The metallographic structure reveals the multicrystallite aggregation of a phase or of a mixture of phases. This aggregation may be described in terms of the number of phases, the morphology of the phases, and the configuration of phases. The microstructure cannot of itself give information on the chemistry of phases or on their crystal structure. Such requirements need the assistance of other measurement systems.

This book will not concern itself with techniques of preparation of a metallic specimen for examination. It is assumed that the reader is familiar with the necessity of creating a flat and highly reflective surface, free of observable physical imperfections and unchanged in

1

nature by the method of preparation. The resolution of phases and grain boundaries is achieved through the action of etching, most commonly by aqueous solutions. The choice of etching reagent is still in an empirical state, but the qualifications of such can be laid down.

Etching is fundamentally a highly anisotropic dissolution process. It is required that an etching solution attack each type of phase present at significantly different rates. In this way each phase is characterized in terms of light reflectivity by a consistent appearance which is somewhere in the spectrum between shiny and black. Not only must the rate of chemical attack be significantly different among the phases present in a structure, but there should also be an observable difference between the various grain or crystallite orientations revealed by a horizontal section. Alternative to this, the rate of etching must be sensitive to small changes in chemistry between the centers of grains and interfaces or boundaries between grains. All of this is meant to imply that the initially flat and uniformly reflective surface must be modified to one which is a series of plateaus, the surface of each having a roughness characteristic of phase identity and the interfaces between represented by sharp, deep, but narrow trenches. A better etchant is one which gives sharper resolution to these differences by local chemical dissolution. Often one etchant is insufficient to distinguish all features. The etchant which makes for sharp contrast between chemically dissimilar phases may be quite incapable of resolving orientation variations in a given phase region.

An alternative means of phase identification is the process whereby the surface is oxidized in chemical solution at a rate characteristic of the phase species. The spectrum of reflectivities in this case is supplanted by a spectrum of colors, whence the term "stain" etching. The color spectrum itself is the product of the relationship between light interference and oxide film thickness, and the structure produced is not fundamentally different from the well-known temper colors encountered in the heat treatment of steel. As with rate of chemical attack, the degrees of stain coloration (or oxidation) must be such as to distinguish variations in one phase type from variations in another. This has not always been adequate because changes in orientation can produce as large differences in coloration as changes in chemistry.

There are still further techniques involving elevated temperature differential oxidation or evaporation. There are also occasions when natural light reflection differences or relief polishing are adequate. The use of polarized light is often valuable where certain phases have optical anisotropy.

It must always be appreciated that metallographic examination constitutes simply a planar section view of a three-dimensional structure. It is not enough to recognize this fact; one must also understand how shape in a three-dimensional construction can degenerate into traces in random planar section. In effect, one must be able by mental-visual skill to recreate from slices of hard-boiled egg, the oblate ellipsoid whence they came.

Our increased ability to interpret the structures which appear before us under the microscope reflects our increased knowledge of the nature of liquid and solid state transformations. At least in a qualitative sense there is good understanding of nucleation and growth processes, of diffusion and diffusionless processes, of the anisotropy of growth, of the influences of temperature and time on transitions from metastability to stability. In such studies, metallography has been a necessary tool. As with many of our tools of study, their use and accomplishments have in turn enhanced their own capability.

Perhaps more than anything else, the understanding of the role of surface or interfacial energy in growth and transformation processes has broadened our power of interpretation. Interfacial energies dominate the transitions from metastable to stable states and the morphologies and configurations of phases which develop.

By developing interpretive skill, one can attempt to rationalize the significance of a new and unfamiliar structure. But this must be done within the context of all available information. Metallographic interpretation is not a self-sufficient process. It must utilize what is available in terms of composition, established phase equilibria, process history, and service experience, as well as what can be made available in terms of physical property measurements, microprobe analysis, and structural studies.

In any system of interpretation, it is the generalizations that are most important. Structures encountered in brasses have counterparts in ferrous and other alloy systems. The transformation of austenite to pearlite in carbon steel is not fundamentally different from the decomposition of the Mg_7Zn_3 phase to Mg terminal solid solution and MgZn. The kinetics may be different but the variations of morphology and the derivative processes are the same.

CHAPTER I

Polycrystalline Structures

Equiaxed, Single Phase, Polycrystalline Grain Structure

The microstructure of Fig. 1.1 portrays a polycrystalline assembly of single phase species. As such it represents a planar cross section of an aggregate of single crystals as in Fig. 1.2 fitting together to fill space completely. The polygonal shapes of the individual grains in the microstructure illustrate versions of idealized solid shapes which concurrently satisfy requirements to fill space completely, conform to topological restrictions, and balance interfacial tensions at intersections.

Any large and representative sample of fitting polygons in two-dimensional arrangement must conform to a topological relationship involving the number of polygons, P, and the number of sides on each polygon, n.

$$\Sigma(6 - n)P_n = 6 *$$

A test of conformity to this rule serves to establish the adequacy of the sample. A similar relationship holds for an assembly of fitting polyhedra of number, B, having polygonal faces, P_n, each having a number of sides, n.

$$\Sigma(6 - n)P_n = 6(B + 1) *$$

Note that in this scheme, there is no necessity for the polygons and polyhedra to be of identical size or shape.

But the grains of a metallurgical material are not simple geometrical arrangements. The bounding faces between crystallites have surface or interfacial energies. Except for low angle and twin relationships, the interfacial energies of randomly oriented grains are all nearly of the same magnitude. This means that along a line of intersection of three adjacent grains, surface tension vectors in each interfacial plane and hence the planes themselves in the immediate vicinity

---

* C. S. Smith, "Metal Interfaces," pp. 65–113. ASM, Metals Park, Novelty, Ohio, 1952.

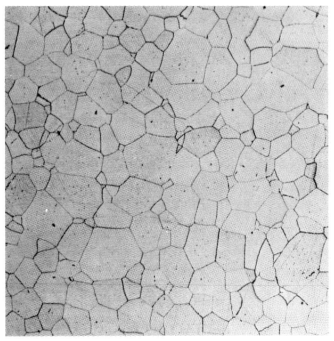

FIG. 1.1. Single phase, equiaxed grain structure of a Mo-W alloy containing 30% W. The planar section reveals polygonal traces with sides varying in number from 3 to 11 of which more than 80% have 4 to 7 sides.
Etchant: 5 gm NaOH, 5 gm $K_3Fe(CN)_6$, 100 ml H_2O. $\times 75$.

of the intersection must be arranged at an angular relationship of 120° to each other. It is generally observed that the statistical distribution of angles between grain boundaries in the annealed, single-phase state peaks at the 120° angular magnitude. However, single real grains are not of identical size and shape so that if the extremities of lines of intersection are forced to conformity by surface tension balances, these lines of intersection must be curved and the polyhedral faces of grains must be curved surfaces. This is apparent in Figs. 1.1 and 1.2.

The inherent curvature of grain boundaries (grain interfaces) is the driving force toward hexagonal grain interfaces and to the growth of grains to equal size. Both of these factors minimize curvature and hence the total energy of the interfacial system.

Topological relationships for space filling polyhedra provide another important relationship, namely, that the function $n/(6 - \bar{n})$ should be as large as possible (where n is the average number of edges on

Fig. 1.2. These are individual grain groupings which parted from an arc cast Ti alloy billet under the blows of a hammer. These fragments have preserved the true bounding facets of the individual grains which belonging to a cast structure are unusually large. Such perfect intercrystalline cleavage is rather rare and is usually associated with a thin intercrystalline film of a low melting, liquid phase whose dihedral angle with the solid is very small. ×1.

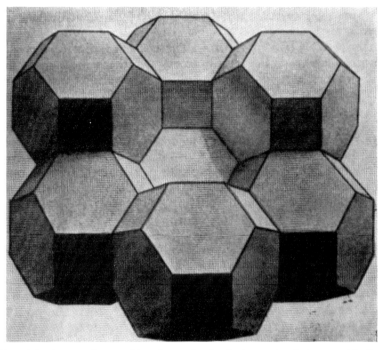

FIG. 1.3. A nested assembly of tetrakaidecahedrons simulating an idealized polycrystalline aggregate. From C. S. Smith, "Metal Interfaces," ASM, Metals Park, Novelty, Ohio, 1952. Drawing by C. S. Barrett.

polyhedra or the average number of sides on the polygonal traces of polyhedra in a micrograph). For this function to be physically real and very large, $\bar{n}$ must be 6 or slightly less. Actually in Fig. 1.1, the 187 grains have an accumulated number of 1056 sides and an average of 5.65 sides per grain.

If we seek a shape which can be packed to fill space completely, a shape whose polygonal faces are in the majority hexagonal (and none with a larger number of sides) and whose edges make angles of 120°, the ideal is the tetrakaidecahedron of Lord Kelvin illustrated in Fig. 1.3. We can assume that grain growth, reduction of grain edge curvature, and the elimination of energetically unfavorable interfaces are processes directed toward approach to this ideal.

Lest it be thought that this discussion is peculiar to metals, Fig. 1.4 is presented to show the same characteristic polygonal structure in a pure crystalline ceramic.

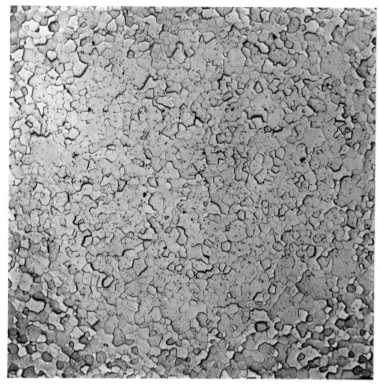

FIG. 1.4. Simple equiaxed grain structure of high density MgO produced by
hot pressing of fine, high purity powder to which has been added a small amount
of CaF$_2$ or LiF$_2$. The fluoride melts in the hot pressing operation providing a
liquid-phase sintering action. The fluoride, however, has sufficient solid solubility
in MgO to disappear ultimately leaving a dense, single phase, dilute solid solution. One could accomplish the same end with tungsten powder to which has
been added a small amount ($\sim$0.5%) of nickel powder and the aggregate hot
pressed or cold pressed and sintered above the melting point of the latter.
Unetched. $\times$100.

Subgrain Boundaries

The individual grains in Fig. 1.5 are subdivided by interlocking
networks of subboundaries which are significantly lighter etching.
Intersections with primary grain boundaries occur frequently at nearly
right angles which indicates that interfacial tensions along these subboundaries are small.

Slip lines in individual grains generated by deformation are seen
to traverse through subgrain boundaries without apparent obstruction

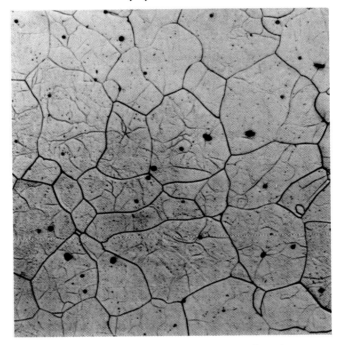

FIG. 1.5. Unalloyed iron (Armco grade). Subgrain boundaries in iron can only be brought out by a prior conditioning heat treatment which permits segregation of interstitial elements to those low angle interfaces. The segregation which is not necessarily actual precipitation increases the rate of etching at the subgrain boundaries. This particular specimen was furnace cooled from 2200°F. At that elevated temperature all prior segregation is erased and the interstitial elements are randomly disposed throughout the grain. During slow cooling, the interstitial elements "condense" on the low angle boundaries. This latter process can be forestalled by quenching from the upper temperature and under such circumstances, the low angle boundaries cannot be revealed by etching.

Etchant: 2 gm picric acid, 2% HNO₃, 98% ethyl alcohol. ×150.

or change in direction as in Fig. 1.6. In contrast, slip lines are stopped by primary grain boundaries or forced to change direction abruptly. From this one can guess that the angular relationships between subgrains are small. From studies of the Laue spots produced in X-ray diffraction patterns, it has been demonstrated clearly that the subgrains of a major grain are actually volume fractions differing in orientation from each other by only a few degrees.

Conditions for the formation of subgrain boundaries are fairly well understood. They are most easily seen in coarse grained specimens which have experienced cold or warm deformation and have been reheated nearly to a recrystallization temperature and cooled there-

FIG. 1.6. This field focuses on the interior of a grain of hafnium in an arc cast ingot with large columnar grains. The macrograins are made up of an elongated cellular arrangement of subgrains. The minor orientation differences between subgrains are revealed by the undisturbed direction of slip bands emanating from the microhardness indentation. The angular relationships between intersecting slip bands suggest that this hexagonal close packed structure develops slip in the prismatic planes rather than the basal planes.
Etchant: 20% HF, 20% HNO₃, 60% glycerin. ×500.

from slowly. The latter step is not implicit in the development of the subgrains themselves but in the response to etching by which they are revealed metallographically. This response to etching seems to be related to the segregation of minor alloy species at subgrain boundaries of trace elements, particularly interstitial elements. Subgrain structures may go unobserved if a species of segregant does not exist which enhances the etching rate of the low angle boundaries.

In the present understanding of low angle or subgrain boundaries, they can be the result of the thermal reorientation of dislocations produced in large numbers and more or less random arrays by small amounts of deformation. Thermal activation permits dislocations of like sign to adopt a minimum energy condition which is a long line. The randomly growing lines intersect to form the network observed.

Subgrain structures appear also in the large grains of cast metals apparently as a result of the freezing process. The main low angle interfaces are usually oriented in the direction of freezing. A line of dislocations of like sign physically constitutes a low angle boundary between slightly disoriented volume fractions of a crystal. The number of dislocations per unit length of line defines the angle of orientation. At high magnifications subgrain boundaries can often be resolved into a series of dots or etch pits which are thought to represent individual dislocations.

The term polyganization was originally used to describe the regular array or lines of etch pits developed in single crystals of certain hexagonal close packed metals by low temperature thermal-mechanical treatment. Subsequently polyganization was found to occur in single crystals of metals other than hexagonal and also to occur in the individual grains of a polycrystalline aggregate.

Subgrain boundaries are thermally very stable but frequently can be apparently erased by thermal-mechanical treatments which produce a fine grained recrystallized structure. Subgrain boundaries often serve as preferred sites of precipitation of a second phase from supersaturated solid solution* and of segregation of solute atoms while still in solid solution.

Metallography of Cold Working

Capacity for plastic deformation distinguishes metals from other engineering materials perhaps more than any other property. Plastic deformation in metals is both a means for constraint to shape and a method of inducing structural, mechanical, and physical property changes. The structural effects observed are mostly rationalized in terms of the crystalline and, in particular, the polycrystalline character of metals.

The primary unit of deformation in individual crystals is slip. Physically, slip is the translation of one portion of a crystal past the remainder on a plane of symmetry which is called a slip plane. Crystal forms usually prefer one or a few types of potential slip planes with specific crystallographic descriptions in terms of Miller indices. In addition to preferred slip planes there are also preferred slip directions. Probably the slip direction is the more important qualification of a slip system, for as in body centered cubic iron the slip direction [111] is invariant but any of the slip planes (110), (123), and (112) can operate.

---

* See section on Precipitation from Solid Solution, Fig. 4.5. (Chapter IV).

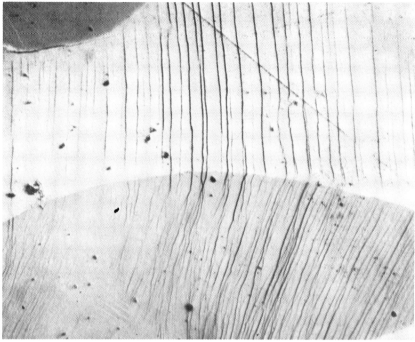

Fig. 1.7. A composite of mild steel fibers and silver was produced by infiltration of a sintered pad of felted fibers with molten Ag. The resulting structure was dense and pore-free, with Ag the major phase. Since Ag and Fe are immiscible there was no alloying action; only wetting of the solid Fe by the liquid Ag occurred. This micrograph represents an analysis of the distribution of plastic strains in the two components of the composite. The gauge length of a tensile specimen was polished to metallographic quality before stretching only a few per cent. The slip line pattern produced is a relief structure which would disappear on repolishing. The micrograph illustrates several points. Slip proceeds simultaneously in both phases. In fact, slip in one phase is clearly related to slip in the other because every slip band in the silver (white) continues with changed direction and character into the steel (dark). This micrograph also illustrates the relative straight and simple character of slip lines in fcc Ag as compared to the more complex and very wavy slip line systems in bcc Fe.

Unetched. ×450.

Not every potential slip plane becomes operative on reaching a critical shear stress. Active slip planes are usually widely separated, the separation distance being temperature dependent. Except in hexagonal close packed metals with exclusively basal slip, slip systems intersect each other. The plastic change in shape of a crystal is the aggregate of slip displacement in a large number of parallel and intersecting slip systems.

FIG. 1.8. This microstructure is the result of an experiment to determine the
distribution of plastic strains in a composite of short length Mo fibers dispersed
throughout a matrix of Cu. The specimen was machined so that it could be
stretched in unaxial tension by small amounts. The surface of the gauge length
f the specimen was polished to metallographic quality. There was no need for
etching to resolve the two constituents of the structure. The relief pattern of the
slip lines tells the story of plastic distribution in the composite. The Mo fibers
showed no disposition to deform as judged by the absence of slip lines. Slip is
totally confined to the Cu matrix. The parallelism and periodicity of slip lines,
intersecting slip, and deformation banding are revealed in this micrograph. On
repolishing and etching the general slip line pattern will disappear because no
orientation changes are involved. However, one can see regions of heavy intersecting slip which has brought curvature to certain families of slip lines
and these of necessity imply the development of orientation variations whose
existence will persist in repolishing and etching.
    Unetched. ×135.

Slip does not involve reorientation about the slip plane so that a
freshly polished and etched surface will not reveal the patterns of
prior slip. Simple slip can only be observed as a relief effect or as a
post-aging effect. The shear displacement on a slip plane which emerges
to a freshly polished surface creates a step along a line trace which is
visible even at low magnifications. Under special conditions of ex-

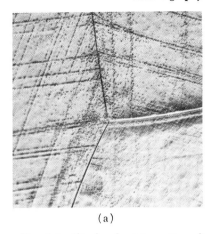
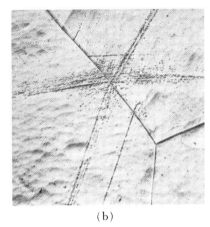

(a)   (b)

FIG. 1.9. Slip bands, intersecting slip (a), and grain-to-grain propagation of slip (b) are illustrated in these two micrographs of Nb containing $\sim$0.01% oxygen. After a small amount of plastic prestrain at room temperature, the specimens were aged at 300°C for 16 hours. During this time, oxygen atoms segregated to the dislocation arrangements within the slip bands. On random section, the free ends of these dislocation lines emerge to the surface and, by virtue of the oxygen segregation, are preferentially attacked by the etching reagent to form a well-defined pit (etch pit).
Etchant: (a) and (b). Electropolished and etched in 10% HF, 90% H_2SO_4. $\times$450.

amination the surface steps can be revealed as an accumulation of smaller unit steps indicating that slip has occurred in a large number of slip planes closely grouped together. For this reason, the traces of slip shown in Fig. 1.7 are usually called slip bands. Repolishing and etching of this same specimen causes the disappearance of most of the families of slip bands. When a multiplicity of slip planes is permissible as with ferritic iron, the slip bands may be wavy in appearance rather than straight, indicating that the slip process is moving from one plane to another on a submicroscopic scale. Wavy slip bands can also be produced as in Fig. 1.8 by multiple intersecting slip.

The highly localized nature of slip becomes more pronounced with increasing degrees of deformation. In the coarse grained specimen shown in Fig. 1.8 the accumulated slip has produced a minor change in orientation of a band of metal within the crystal. These deformation bands have reoriented sufficiently to allow resolution after etching.

The occurrence of slip in individual grains and from grain to grain can be observed under certain conditions by the process of dislocation decoration. An example is shown in Fig. 1.9. Dislocation decoration

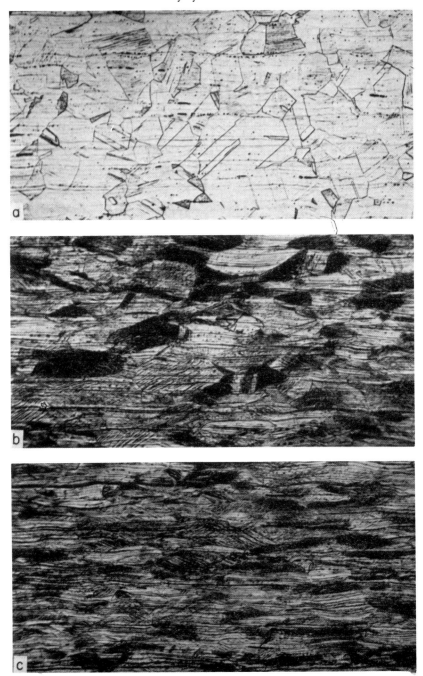

FIG. 1.10a,b,c

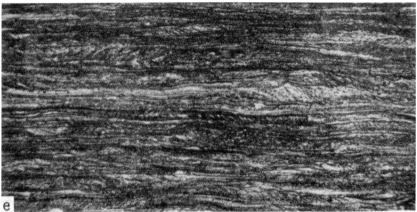

FIG. 1.10. This series of microstructures illustrates the changes accompanying increasing degrees of cold reduction applied to the initially recrystallized, equiaxed condition for an austenitic stainless steel (18% Cr–8% Ni). Key: (a), 0%; (b), 20%; (c), 40%; (d), 60%; and (e), 80% cold reductions. In the 20, 40, and 60% cold reductions there is a gradual distortion of the individual grains and a gradual loss in etching contrast and definition of grain boundaries. In fact the grain boundary is ceasing to exist as a zone distinct in structure and chemical reactivity from the grain center. The grains themselves are so heavily populated with intersecting deformation bands that crystalline disregistry permeates the whole body. After 80% cold reduction, the polycrystalline appearance of the metal is gone and in fact this nearly so, for in developing a deformation texture or preferred orientation, the specimen is degenerating to only a few distinguishable orientations.

The recrystallized state as it appears here reflects the poor quality of this particular specimen. There are stringer inclusions in large amount. The parallel ghost lines represent some sort of banding inherited from ingot segregation. In this case it is probably carbon banding. The etching effect derives from localized precipitation of carbides because the effect disappears when the specimen is reheated to 1550°F for 15 minutes and water quenched.

Etchant: (a)–(e) 5 gm CuCl₃, 10% HCl, 90% ethyl alcohol. ×135.

depends on the ability of certain solute elements at very dilute concentrations to segregate to the individual dislocation loops lying in the slip plane. This usually requires a low temperature postheat treatment sufficient to enhance mobility of the solute atoms but not to relax and cancel the existing dislocation array.

One distinguishes between cold and hot deformation not by the level of temperature involved but by whether the temperature is below or above the range for short-time recrystallization. Thus refractory metals can be cold worked many hundreds of degrees above room temperature, whereas room ambient is in the hot working range for very pure lead. The preservation of crystal distortion as seen by postmetallographic examination is a primary basis for judgment of the kind of deformation experienced.

Figure 1.10 shows a typical sequence of the distortions undergone by the grains of a polycrystalline aggregate. The individual grains become elongated in the directions of principal strain. A gradual change in etching behavior occurs with increasing cold reduction. This also coincides with the gradual development of a preferred orientation or texture which signifies that the grains are not only elongating but their orientations with respect to some arbitrary direction are changing from random to preferred. The arbitrary directions are those of the principal strains, and the preferred orientations are described in terms of the crystallographic directions parallel to the directions of principal strain and the crystallographic planes parallel to important external surfaces or symmetric planes of section. This gradual loss of large orientation differences between adjacent grains tends to reduce etching definition of grain boundaries. In metals which have undergone cold reductions of 50% or more, one cannot usually follow the complete outline of any single grain.

The microstructural changes attendant on cold working of two-phase structures depend very much on the mechanical properties of the minor phase. In fact the actual ability of the alloy to sustain deformation is critically dependent on the nature, amount, and distribution of minor phases. When the minor phase is soft and ductile, it will elongate with the grains of the major phase, and since interfacial bonds are strong the degree of deformation will appear to be the same. A brittle minor phase, if permitting deformation at all without cracking, will simply string out along the directions of principal strains. When the brittle phase is platelike in shape and its long axes are inclined to the directions of principal strains, fragmentation may also occur.

There is an intermediate condition where isolated grains of a normally brittle metal behave in a ductile manner. A case in point is illustrated in Fig. 1.11. The structure shows tungsten grains enveloped by 10–20% by volume of a minor phase which is a nickel-rich solid solution containing iron and tungsten (face centered cubic). Polycrystalline tungsten at room temperature is quite brittle, yet in this two-phase structure the grains of tungsten demonstrate extensive cold ductility.

The phenomenon is not altogether mysterious for it has been known for many years that single crystals of tungsten are ductile. The structure in Fig. 1.11 can be regarded as an assembly of single crystals of tungsten whose interface with a ductile enveloping phase will permit the passage of slip bands. However, it must not be inferred that envelopment by any ductile phase will serve the same purpose. An envelope of copper, for example, will not permit any appreciable cold work. Studies on this type of structure indicate that there must be an approximate equivalence in the flow stress of the particulate and envelope phases. Under such conditions, other room temperature brittle metals such as Cr, Mo, and Be can be made to exhibit ductility. It will be noted in Fig. 1.11 that the etching definition of the tungsten-envelope phase interfaces is not reduced even with large degrees of cold work. This is because large chemical differences are preserved across the interfaces.

Certain metals exhibit a discontinuous yield effect in which the stress to initiate yield is greater than the stress to sustain it. This produces a steplike shear displacement on the surface of the metal not unlike the relief effect produced by a slip band but with the significant distinction that the line or trace of displacement runs continuously through many grains. These Lüders bands are most commonly associated with mild steel but are found in some aluminum (Fig. 1.12) and copper alloys also. As with slip bands they are most readily observed on surfaces freshly polished before deformation, and, on repolishing and etching, the effect disappears. This signifies that Lüders bands involve simple shear displacement and no reorientation on either side of the band. Lüders bands appear to follow the directions of principal macroshear strains and are often useful in revealing these. The complex shear strain patterns at the base of a Charpy V-notch impact specimen are revealed in Fig. 1.13 by another process.

Mechanical twinning is a phenomenon associated with cold deformation involving the cooperative reorientation of a plate-shaped zone of a crystal under the action of shear stresses. It may precede, follow, or

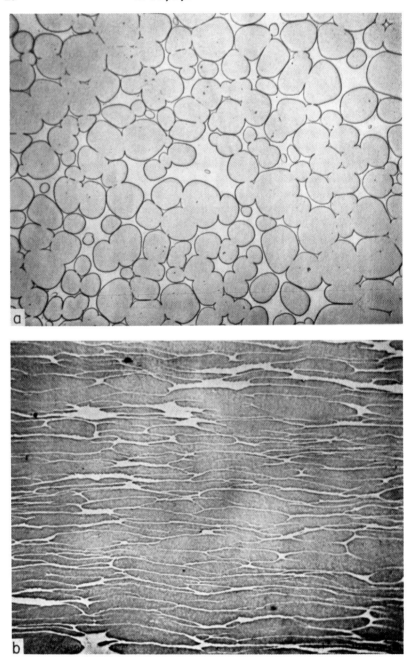

Fig. 1.11

occur concurrently with slip. The rate of formation of mechanical twins is very fast and is often accompanied by audible release of elastic stress waves. Tin "cry" is just this effect.

The twin relationship represents two volumes of the same phase bounded by a plane interface across which a specific orientation relationship exists. The orientation relationship is most easily visualized as a mirror image effect produced by certain crystallographic directions and planes of one component of the twin and its corresponding directions and planes in the other component of the twin.° The twin interface is a low energy boundary and therefore maintains a planar character. A twin appears as a plate bounded by two parallel twin interfaces. Since lateral growth involves distortion of the twin planar interfaces with consequent high absorption of energy, it does not easily occur. Accordingly twins are very thin plates (Fig. 1.14). Since the twin involves a large change in orientation from the original matrix,

° See C. S. Barrett and T. B. Massalski, "Structure of Metals" (3rd ed.), pp. 406–416. McGraw-Hill, New York, 1966.

FIG. 1.11. This material has a nominal composition of 90% W, 7% Ni, and 3% Fe. Yet it is in reality a composite of two materials, one of which is essentially pure W—the spheroidal grains (a)—and one of which is a binary alloy of 70% Ni and 30% Fe with perhaps only a few per cent W in solid solution—the envelope phase. The composite structure can be cold rolled great amounts as illustrated in (b) for the condition of 90% reduction in strip thickness. Moreover as can readily be seen, both phases participate in the deformation process.

Ordinarily, polycrystalline W is brittle at room temperature, yet in this instance W is clearly highly deformable. One can draw the conclusion that the W-W grain boundaries are somehow the obstacle to the propagation of slip and that when these are removed or replaced by certain other interfaces, deformation by slip propagation becomes possible. Actually it has been known for many years that single crystals of tungsten are quite ductile.

Some W/W interfaces exist in this structure but cracks, if they do generate there, cannot propagate more than to a neighbor grain of W. This becomes the same sort of conditions as in hot working of structures containing brittle compound networks. The cracks are stopped by the envelope phase.

The spheroidal form of the W grains prior to cold rolling is the result of a growth-from-the-melt process. The specimen as a physical mixture of elemental powders is sintered at a temperature above the melting point of the 70% Ni–30% Fe alloy. This melt is capable of dissolving large amounts of W so that at equilibrium at this temperature (~1500°C) the melt occupies perhaps 50% by volume of the structure. On cooling, the melt rejects W which deposits onto the existing grains of that element. So low is the interfacial energy of the W-melt interface that there are no preferred growth directions and the crystallization structure is spheroidal rather than dendritic.

(a) Unetched. ×250.

(b) Etchant: 5 gm NaOH, 5 gm $K_3Fe(CN)_6$, 100 ml H_2O. ×250.

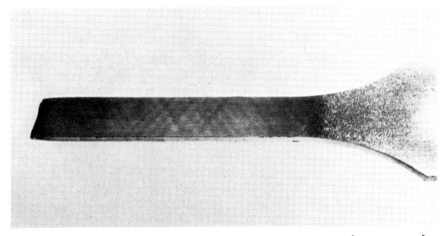

FIG. 1.12. These Lüders bands have been produced in a tensile specimen of an Al–4% Mg alloy. They represent macroshear displacements which have occurred in the first few per cent elongation at the yield point. The individual shear displacements represent the accumulated and cooperative slip by all the grains traversed by the Lüders band. This phenomenon is associated with a condition where the stress to propagate slip from grain to grain is less than the stress to initiate such action. The elastic energy stored at the time of slip initiation in the stressing system is enough to force propagation of slip across all grains in the plane of maximum shear stress.

This effect is more commonly associated with the influence of dissolved nitrogen and carbon in mild steel, where theory argues convincingly that these elements tend to segregate to the sources of slip and increase the binding energy of the basic units of slip (dislocations). Having broken these bonds and traveled out of range of interaction, impediment to slip is greatly diminished and it can proceed at a much lower shear stress.

Even though the diffusion mobility of Mg in Al is much lower than C in Fe, the atomic population is orders of magnitude higher. Any Al atom is probably never more than two atomic distances away from a Mg atom. Because of this, re-locking of slip bands occurs rapidly at room temperature as evidenced by serrations in the whole stress-strain curve.

For some practical illustrations of stretcher strain markings see C. J. Smith, *Metal Ind.* (*London*) 98, 499–501 (1961).

Unetched. ×1.

the traces of twin bands are easily revealed by etching. With subsequent deformation, twin bands increase in orientation multiplicity and old ones become distorted. In losing their straight band symmetry they can appear similar to wavy deformation bands.

Mechanical twinning is common in noncubic metals but under conditions of very low temperature and high strain rates, it may

FIG. 1.13. This represents a Charpy V-notch impact specimen of mild steel which has been struck with less energy than is necessary to completely fracture. A small crack has just started at the base of the machined notch. After the reduced impact blow, the specimen was aged at 250°C for 30 minutes. The view shown is of a mid-section. As a result of the cold deformation and the aging treatment carbon segregates either as dislocation assemblies or as an actual precipitate. This segregation is reflected in the selective etching by the cupric chloride solution. The selective etching reveals the patterns of shear strain about the root of the notch and at the opposite face. There does not seem to be any obvious explanation for the formation of the alternate light and dark etched bands but the effect is real and meaningful. See also A. P. Green and B. B. Hundy, *J. Mech. Phys. Solids* **4**, 128–144 (1956).
Etchant: 45 gm $CuCl_3$, 180 ml HCl, 100 ml H_2O. ×4.

appear in most body centered cubic metals. In ferritic iron, mechanical twins are called Neumann bands. Mechanical twins are often found in the vicinity of the path of brittle fracture, which has excited a controversy for many years as to whether mechanical twinning nucleates brittle fracture or vice versa. With prior knowledge of twinning conditions for a particular metal, metallographic examination can be a valuable indicator of the existence of low temperature mechanical or thermal stressing and of the nature of the conditions for an existing fracture in post mortem study.

Twin relationships and twin plates can form as a product in crystallization and recrystallization processes. They have the same general appearance as mechanical twins except that the bands are

FIG. 1.14. With large additions of Re to the solid solutions based on W or Mo, there occurs a strong propensity for mechanical twinning as illustrated here for a specimen of arc cast W–30% Re. The twins themselves were produced in the cutting or breaking operation to prepare a piece for microexamination. Repeated polishing will not remove them. Only if the original cast button were carefully polished could a twin-free structure be observed. The occurrence of a strong propensity for mechanical twinning coincides also with a considerable increase in warm workability.

Etchant: 20% HF, 20% HNO₃, 60% glycerin. ×250.

usually much wider or thicker. Among the cubic metals, recrystallization twins are common and mechanical twins are very uncommon in the face centered cubic varieties and the converse is true in the body centered cubic varieties. Occasionally both growth twins and mechanical twins occur in the same structure (see for example Fig. 3.24).

Certain thermally stable but thermodynamically unstable high temperature phases can be induced to transform martensitically by cold work. The process is not unlike mechanical twinning, with the major distinction being the generation of a crystallographically new identity. A case in point is the generation of body centered cubic martensite in room-temperature-retained austenite of certain stainless steels. Figure 1.15 illustrates the results of severe cold working on the

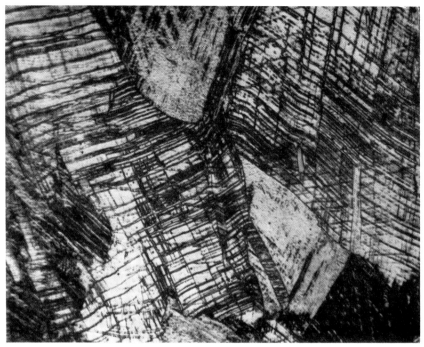

FIG. 1.15. Common stainless steel (18% Cr, 8% Ni) is single phase austenite under normal conditions of working and annealing. It is, however, metastable at subzero temperatures with respect to the formation of a martensite isotypic with ferrite. The transformation can be brought on by deformation at low temperatures. The specimen shown was cold rolled 30% at —65°C, resulting in a great increase in strength. With successive degrees of deformation, more and more martensite is formed. In the microstructure shown, the bands of martensite which formed with small strains have become bent and distorted by the continued deformation. Without other information this type of structure would be indistinguishable from mechanical twinning produced by rolling to the same degree. Etchant: 5 gm CuCl₃, 10% HCl, 90% ethyl alcohol. ×900.

formation of martensite bands, their intersection, displacement, and bending produced by later generations of martensite, and general slip in the two phases. Deformation-triggered martensite usually develops in thin bands which are not obviously distinguishable from twins except by X-ray diffraction or by polarized light if one or other but not both phases are optically anisotropic.

Recrystallization and Grain Growth

Recrystallization pertains to the nucleation and growth of new, strain-free crystals in an existing assembly of grains which have been

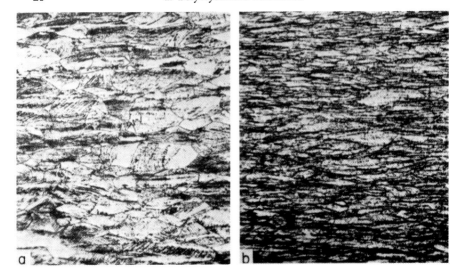

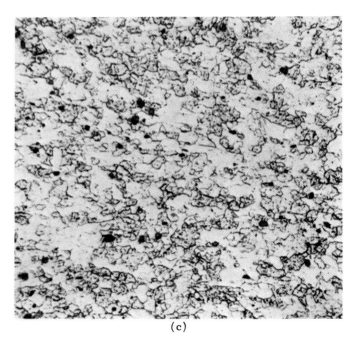

(c)

FIG. 1.16a,b,c

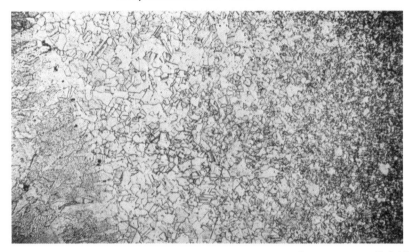

FIG. 1.17. The heat-affected zone of a weldment encapsulates into a small field of view all of the recrystallization and grain coarsening events which are likely to occur in a cold worked metal. This micrograph is a view of the heat-affected zone of butt welded Cu sheet which had been joined by a W-inert gas fusion welding operation. At either extremity are the cast metal of the weld and the original cold worked structure of the adjoining metal. The recrystallization and grain coarsening all took place in a fraction of a minute. Note the line of gas porosity at the juncture of the original interface between solid and molten metal. This demonstrates that gas as well as solids prefer to nucleate heterogeneously.

Etchant: 33% H_2O_2, 33% NH_4OH, 34% H_2O. $\times 35$.

FIG. 1.16. Micrographs (a) and (b), of 18% Cr–8% Ni austenite stainless steel in an incomplete state of recrystallization, are to be compared with Fig. 1.10 which illustrates the prior states of 20% cold reduction [(a) and Fig. 1.10(b)] and 80% cold reduction [(b) and Fig. 1.10(e)]. Both (a) and (b) were annealed at 1550°F for 5 minutes, and both have suffered a large drop in hardness and are nearly completely soft ($\sim$180–200 DPH). Yet by simple examination, only (b) perhaps would indicate a prior annealing history. Most of the microstructural changes—sharpening of new interfaces—will occur in the small hardness change from 206–233 DPH to 180–200 DPH. Such an instance for partially recrystallized copper is shown in micrograph (c). Hardness testing therefore is more sensitive to the progress of recrystallization in the initial stages, and metallography is more indicative of changes in the latter stages.

Etchant: (a) and (b). 5 gm $CuCl_3$, 10% HCl, 90% ethyl alcohol used electrolytically. $\times 135$. (c) 33% H_2O, 33% NH_4OH, 34% H_2O. $\times 250$.

Initial hardness: (a) 397 DPH, (b) 412 DPH. Annealed hardness: (a) 206 DPH, (b) 233 DPH.

rendered thermally unstable by deformation. Strain-free nuclei grow radially at roughly isotropic rates. The growth occurs at the expense of the strained polycrystalline matrix, and when recrystallization is complete, all of the original matrix has been consumed by the new grain systems. The impingement of adjacent new grains generates the cellular network of interfaces which examination in planar section reveals as a polygonal network of grain boundaries.

The progress of recrystallization can be followed metallographically more easily in heavily cold worked metal because the new grains etch sharply and stand in contrast to the ill-defined, more darkly etching structure of the strained matrix. Distinguishing between strained and recrystallized material in structures which have received small cold reductions ($<30\%$) is much more difficult. In such circumstances, it is perhaps better to follow recrystallization by some physical or mechanical property such as hardness. Some of these points are illustrated in Fig. 1.16.

The rate of recrystallization is governed primarily by two factors—temperature, and degree of prior cold deformation. The term "recrystallization temperature" usually signifies that temperature at which the process is rapid and achieves completion in a matter of minutes (e.g. 30–60). One must not forget, however, that complete recrystallization can be attained by prolonged exposure to significantly lower temperatures. In general, larger degrees of prior cold work lead to lower recrystallization temperatures. The heat-affected zone about a fusion weld between sheets of cold worked metal (Fig. 1.17) illustrates the influence of temperature on recrystallization and grain growth in one composite picture. In this circumstance, the observed recrystallization is the product of transient temperature surges of less than a few minutes and mostly of seconds duration.

Both the recrystallization temperature and the minimum recrystallized grain size depend on the degree of prior reduction. The higher degrees of prior cold work lead to finer recrystallized grain size but more correctly, the penultimate recrystallized grain size depends in complex fashion on the original grain size, degree of cold work applied, and the temperature of final anneal. The recrystallized grain size is most critically dependent on pre-strains in the range 1–10%. Within this range of cold work, recrystallization can lead to abnormally large grain sizes. This phenomenon provided one of the early methods of producing single crystals large enough to cut out individually for study of the orientation dependence of mechanical and physical prop-

FIG. 1.18. With most pure metals and some alloys there exists a critical strain which upon recrystallization causes the formation of extremely large grains. This critical strain is of the order of 1–10% and the optimum recrystallization temperature for most exaggerated grain coarsening is at the low limit of the range. Many times the critical strain is introduced unwittingly by bending, indenting, or shearing an annealed metal. In the present case, a specimen of zinc in the recrystallized form was cut by hack saw from a larger piece. The heating involved in plastic mounting of the metallographic specimen was sufficient to induce recrystallization in the slightly cold worked metal adjacent to the saw cut edge with the resultant abnormal grain growth. There is a sharp change in grain size to that characteristic of the remainder of the specimen demonstrating the very localized nature of the small plastic strains. Abnormal grain growth can sometimes be encountered in metals which have experienced small degrees of hot work at temperatures near the lower limit of the recrystallization range.

Etchant: 10% HCl, 90% H_2O. ×50 PL.

erties. Figure 1.18 illustrates a recrystallized structure with abnormal grain growth. The abnormal grain size can also result from hot working in the temperature range of recrystallization. Such circumstances can, for example, lead to circumferential bands of abnormally coarse grains in extrusions.

The recrystallization temperature is usually modified substantially by alloy additions. Certain elements can be effective in very small

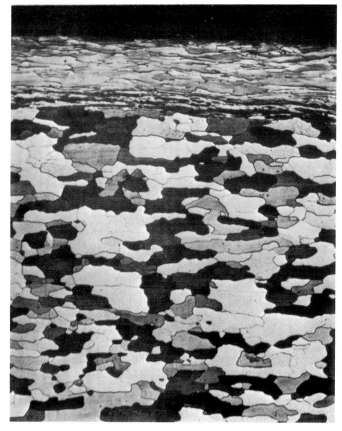

FIG. 1.19. Microstructure of a Mo alloy (0.5% Ti, 0.07% Zr, and 0.019% C) taken near the surface of $\frac{1}{16}$ in. sheet. The specimen has been annealed for 15 minutes at 2650°F which is sufficient to provide for complete recrystallization of the main body of the sheet. The surface, however, during the thermal-mechanical history of processing has acquired a higher concentration of oxygen and nitrogen which has significantly raised the recrystallization temperature so that the initial heavily cold worked structure is completely retained. The elongated character of the recrystallized grains is typical of the behavior of a material containing a finely dispersed phase. In this case, the Ti, Zr, and C alloying constituents are largely bound up as complex oxy-nitro-carbides which are so finely distributed that they can only be seen under the electron microscope at perhaps 10,000 magnification. During the working history these are strung out in sheets in the planes of principal strain providing restraint to grain growth in directions perpendicular to the thickness of the sheet. (Figure courtesy of M. Semchyshen and R. Q. Barr, Climax Molybdenum Co. of Michigan.)
Etchant: Electropolished and etched in NaOH + $K_3Fe(CN)_6$. ×250.

amounts. For example, as little as 0.1% silver can raise the recrystallization temperature of copper by about 100°C. Figure 1.19 illustrates the influence of atmospheric contamination of the surface of molybdenum sheet on the recrystallization behavior. It is thought that the strong effect of very dilute solid solutions is related to the preferential segregation of certain elements to the grain boundaries and subgrain boundaries rendering them less mobile.* Since the recrystallization process is the migration and elimination of these interfaces, anything which impedes their motion raises the recrystallization temperature.

Submicroscopic dispersions which resist grain coarsening are very effective in raising the recrystallization temperature. For example, the complex carbides or nitrocarbides in the high hot strength alloy Mo + 0.5% Ti + 0.05% Zr + 0.05% C prevent recrystallization for prolonged times of exposure to 1480°C, whereas pure molybdenum recrystallizes below 1100°C. Coarser dispersions such as iron in brass restrict grain growth but have little influence on the recrystallization temperature.

The process of grain coarsening is not always separable from recrystallization. Under the circumstances that a fine-grained recrystallized structure is heated to some temperature higher than previously experienced, grain coarsening can be expected. In a simple, single phase structure free of inclusions the grain size is a monotonic function of the peak temperature of heating. In thin sheet, grain growth may be restricted, however, by the free surface. Except under unusual circumstances involving preferred orientations, the grain size is held to average diameters of no more than two or three times the sheet thickness.

Systematic distribution of a dispersed phase is very effective in restraining grain growth. While the particle size of dispersants for restraint to recrystallization need to be of submicroscopic dimensions, much coarser particles are amply effective on grain growth inhibition. Figure 1.20 (a,b,c) illustrates the abrupt coarsening of an austenitic grain size as the dispersed carbides dissolve. One encounters a similar phenomenon in aluminum-killed mild steel at the temperature at which aluminum nitrides dissolve in austenite, although in this case, the dispersed phase is unresolvable by normal optical microscopy.

Recrystallized grains are usually equiaxed, but when existing insoluble inclusions have been strung out by hot or cold working into

* For distribution of silver in the grain boundaries of copper, see "Imperfections in Nearly Perfect Crystals," p. 469. Wiley, New York, 1952.

Fig. 1.20a,b,c

closely spaced planar bands the grains recrystallizing between them will assume an elongated geometry (see Fig. 1.21).

Twins invariably occur in the recrystallized structures of certain face centered cubic metals, notably in copper-base alloys. On the other hand, some face centered cubic metals such as aluminum show no propensity for annealing twins. As in mechanical twinning, the twin bears a very specific orientation relationship to the remaining crystal. Each twin is wholly contained in one crystal and usually but not invariably traverses the grain. The average number of twins per grain seems to depend on alloy content. Twins as illustrated in Figs. 2.4 and 2.16 appear as bands in the microstructure, the long or planar faces being absolutely parallel. Annealing twins are invariably much wider (or thicker) than mechanical twins. The twin boundary with the matrix crystal is a low energy interface as indicated by the almost perpendicular intersection with ordinary grain boundaries even after prolonged annealing.*

Not every banded structure identifies annealing twins. Certain colonies of precipitate can appear as parallel bands traversing a crystal and maintain this configuration even after protracted annealing (see Fig. 4.14). These also correspond to low energy interfaces but with another phase species. Such structures are usually erasable by re-solution treatment and quenching. Otherwise the distinction can be made with the use of X-ray diffraction.

* See for example, "Imperfections in Nearly Perfect Crystals," p. 375. Wiley, New York, 1952.

FIG. 1.20. The existence of an undissolved phase in finely divided form can effectively inhibit grain growth at lower temperatures. At some temperature where the dispersed phase dissolves, the grain size can abruptly increase. Even though the dispersed phase does not reduce appreciably in amount, restraint to grain growth often relaxes abruptly at some temperature below the melting point. Again there is a discontinuous burst of grain coarsening over a narrow temperature range. The case shown here illustrates this point. The structures are of an M2 high speed steel austenitized at 2200°F (a), 2233°F (b), and 2267°F (c) successively. The quenched product is largely martensite but sufficient austenite is retained to show the pattern of prior austenite grain boundaries. The austenite grain size is essentially constant at all temperatures up to 2200°F. Thereafter in the short span of 67°F there is a burst of grain coarsening. Part of this is due to progressive solution of the dispersed carbides and part to the weakened restraint of the carbides to grain growth. M2 high speed steel has nominal composition of 0.85% C, 4% Cr, 2% V, 6.25% W, 5% Mo.

Etchant: (a)–(c) 5 gm $CuCl_3$, 10% HCl, 90% ethyl alcohol. ×375.

FIG. 1.21. Although precautions of degassing and fluxing are taken in the commercial melting of Al alloys, the cleansed and purified metal is finally poured in the open air. Films of oxide formed over the molten stream and residual from the melting operation are carried into the ingot mold and become incorporated in the solidified ingot. In subsequent rolling they are spread out in discontinuous planar films, too thin to be resolved metallographically. Their presence, however, is recognizable through the directional restraints to grain growth. In the microstructure of commercial Al 2024 alloy, the grains although fully recrystallized retain the elongated geometry characteristic of a cold worked structure. This is because the planar oxide films have restricted grain growth in the directions perpendicular to these planes.

Etchant: 10% Fluoboric acid, 90% used electrolytically. ×68 PL.

Metallography of Hot Working

We shall define "hot working" simply as deformation produced at a strain rate and temperature where recrystallization of the grains can keep pace with their distortion. Accordingly we expect to see the microstructure of a hot worked metal show the plastic phases in a recrystallized or equiaxed form with a uniform grain size. Since recrystallization is the boundary condition for the term "hot" in hot working, the actual temperatures for hot working may vary widely—from room temperature for lead to 1500°C for tungsten.

The proper hot working temperature for a metal may be significantly higher than the generally quoted recrystallization temperature. This is because recrystallization for annealing purposes is usually gauged in terms of minutes of isothermal annealing time (60 minutes is a common check point), whereas the times involved either in the total period of deformation or between successive stages of deformation are of the order of seconds.

Certain prior conditions in the cast structure can conspire to prevent the uniform grain size, homogeneity, and equiaxed grain shape which is the ideal outcome of a hot working operation. During hot working, interdendritic oxides and other insoluble phases can be drawn out into semicontinuous sheets separating grain groupings. These place strong directional restraints to growth during recrystallization with the outcome that the structure contains elongated grains which are, in fact, fully soft and free of plastic strain energy (see Fig. 1.21).

Macrosegregation in a cast ingot represents macroregions of gross composition differences. Homogenization annealing prior to hot working will tend to eliminate these. Of what is left, hot working itself by distortion of established composition gradients will hasten homogenization. If both of these operations are inadequate, the regions of differing chemical composition will, as in the case of inclusions, be distorted into planar strips in the hot rolled or forged structure. The planar strip arrangement or "banding" is revealed by etching in any one of several ways. Sometimes, though rarely, when the segregation pattern constitutes variations within the solubility limit of a given phase, the etching intensity is sensitive to variations in composition. More often banding is revealed by nonuniform precipitation or by nonuniform annealed grain size. In either case the arrangement of the nonuniformity will follow the principal directions of hot working. In Fig. 1.22 the banding is revealed by the concentration of pearlite.

Hot working in the temperature range of active transformation has the effect of greatly increasing the rate of the transformation and of the rate at which the transformation products assume shapes governed by minimum interfacial energy rather than growth considerations. A case in point is illustrated in Fig. 1.23 which is typical of titanium, zirconium alloys, and $(\alpha + \beta)$ brasses, among others. The equiaxed form of the distributed α phase is very characteristic of prior hot working in the $(\alpha + \beta)$ phase field. It would take many hours of annealing to induce a Widmanstätten pattern to assume a comparable form. This is illustrated elsewhere (Fig. 4.13).

Many alloyed materials in their cast state possess continuous or semicontinuous dendritic networks of intermetallic compounds. The

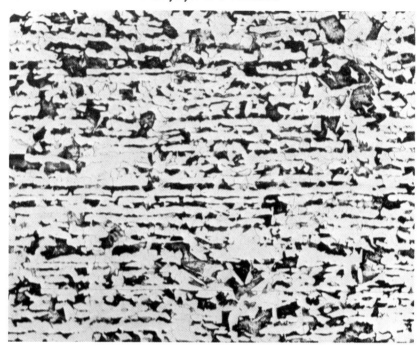

F IG. 1.22. Banding in steel signifies the retention of some form of alloy segregation—either intentional or tramp element species. Carbon segregation on the
"cored" dendrite scale usually disappears during soaking and hot working operations. This steel is a plain carbon type with a nominal level of 0.2% C. The banding
takes the form of alternating elongated regions of all-pearlite and all-ferrite. This
structure is the by-product of a composition banding probably involving high and low
concentrations of phosphorus or manganese. These cause a form of uphill diffusion of
carbon in austenite to or away from the phosphorus-rich areas. The same effect is
produced by copper-rich gradients in austenite (see Fig. 5.8). On cooling, the carbon-
rich regions of austenite transform to an almost all pearlitic structure.

Alloy banding can appear in different forms depending on their influence on
transformation kinetics. Thus there can be alternate zones of coarse and fine
pearlite or pearlite and bainite.

Apart from the directional forming properties of banded steel, there are other
real manufacturing hazards associated with welding and heat treatment followed
by machining operations. The generation of local zones of hard martensite in a
steel, whose nominal carbon and alloy contents would dictate otherwise, can lead
to all kinds of production and operational failures.

Etchant: 2% HNO_3, 98% ethyl alcohol. ×180.

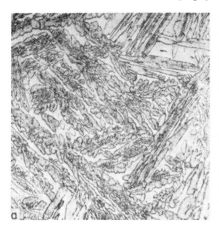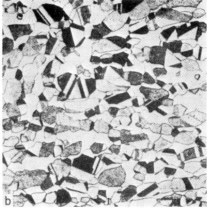

FIG. 1.23. When an alloy is hot worked at the same time as a phase is precipitating, the normal distribution pattern and particle morphology are radically changed. This is Muntz metal—60% Cu, 40% Zn. Micrograph (a) is typical of the Widmanstätten arrangement of α plates in a β matrix produced by cooling from the β phase field. The specimen in micrograph (b) was hot worked at least 50% while it was cooling through the $(\alpha + \beta)$ phase field. The normal precipitation pattern is replaced by what is the closest approximation of an equiaxed grain structure which a two-phase specimen can assume. The difference is in the average dihedral angles at three-grain junctions. This structure could also be attained by reheating the specimen of micrograph (a) back into the $(\alpha + \beta)$ field and hot working an equivalent degree.
Etchant: 33% H_2O_2, 33% NH_4OH, 34% H_2O. ×175.

intermetallic compound, particularly if it is a carbide or other interstitial phase, is itself brittle at all temperatures. Therefore, under plastic deformation, the compound envelopes are cracked repeatedly. Whether the metal fractures depends on the crack-propagating character of the surrounding phases. Usually at low temperatures such cracks can propagate and the cast metal is adjudged brittle at room temperature. Frequently, however, at some elevated temperature, fragmentation of the compound envelopes can proceed without propagation of cracks through the whole structure. This may be achieved with a proper stress state, i.e., extrusion. As illustrated in Fig. 1.24, the compound networks, during and as a result of hot working, transpose to nodular distributions which are far less restrictive to the deformations of the individual ductile-phase grains. Thus by hot working, a cast structure brittle at room temperature may be induced to act in a ductile fashion. As in the case of the α phase in titanium and zirconium alloys, the coagulation of carbides during hot working

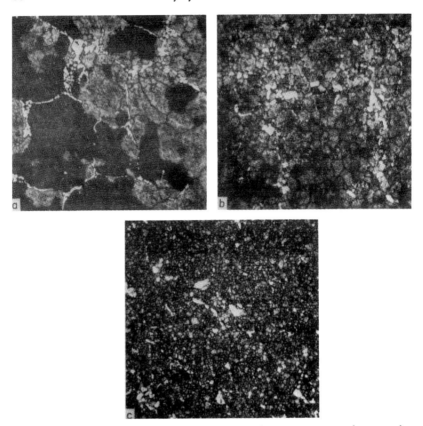

Fig. 1.24. These three micrographs represent three points very close together in the front end of a hot extruded billet of cast high speed steel (5 in. diameter). The actual front end of the 1-in. diameter extrusion is represented by (a). This material was originally directly at the die port and, with the advance of the ram, was pushed out without any hot working. The structure therefore represents the initial configuration of eutectic alloy carbide networks about the arms of austenitic dendrites. A fully hot worked structure is represented by (c) which was taken only 3 in. from (a) and (b) shows a partially worked structure taken at a mid-position about 1½ in. from (a). Comparison of the three micrographs illustrates the influence of hot working on a cast structure of this type. It is clear that hot working has performed the following functions in this case:

(1) eliminated the network arrangement of the massive eutectic carbides,

(2) refined the austenite grain size by about an order of magnitude,

(3) increased the rate of solution of the massive carbides into the austenite matrix since, in the present case, the amount of eutectic exceeds equilibrium requirements.

Etchant: (a) (b) (c). 5 gm CuCl₃, 10% HCl, 90% ethyl alcohol. ×325.

proceeds much faster than would be accomplished by simple annealing. The mechanism for this is simply the continual distortion of existing and forming composition gradients. All forms of homogenization and phase coarsening involve diffusion over microdistances. In any individual diffusion process the rate of diffusion is fastest initially because the surrounding medium is most depleted of the diffusing species. The establishment of a concentration gradient slows down the diffusion transport process, but the hot working by physically distorting the concentration gradients perpetuates the initial transport rate. For example, a 50% reduction in thickness reduces by one-half the distance between composition extremes (in a gradient oriented in the thickness direction) and increases the rate of diffusion by a factor of four.

Working does nothing to reduce the problems of chemistry variations from one end of a billet to the other or from the center of a billet to the outside. These represent macrodistances which cannot be overcome by diffusion action.

Inclusions

The term "inclusion" broadly applies to extraneous particulate matter disposed throughout the normal structure of a metal—more often randomly but, in certain circumstances, systematically disposed. These extraneous materials may have clearly defined chemical compositions and crystal structures such as simple oxides, carbides, and sulfides, or they may be ill-defined such as glassy slags, fluxes, and brick fragments. Even incompletely dissolved metallic alloy additions such as the tungsten inclusion in Fig. 1.25 come within this context.

In general, inclusions are detrimental to the properties of a metal. They impart directionality of mechanical properties to wrought metals, poor quality to machined surfaces, reduced fatigue strength, excessive abrasion to cold working dies, and poor formability to sheet, to name only a few metallurgical ailments. In certain instances they can usefully impart increased hot strength (by internal oxidation), restrained grain growth (by deoxidation), and improved machinability (as in resulfurized steel).

Inclusions most commonly originate in melting and pouring operations. They can also be incorporated in poorly controlled hot working by which surface scale and oxidized cracks are folded into the body of the metal.

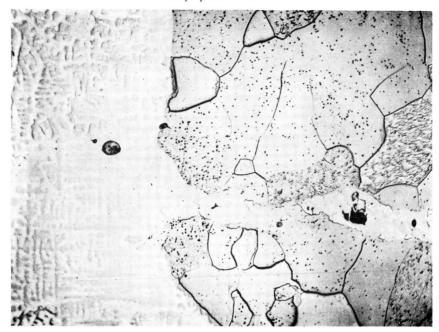

FIG. 1.25. Tungsten is very difficult to add to a melt because of its reluctance to dissolve. When dissolution has been incomplete, examination of the casting reveals inclusions of tungsten. The present example was taken from a small ingot whose intended composition was 65% W, 18% Cr, and 17% Mo. The recrystallized grain structure of the undissolved tungsten particle is contrasted against the coring pattern of the solidified surrounding melt. In some instances (but not this particular case), intergranular penetration of liquid into the solid can accelerate dissolution by essentially rendering large polycrystalline pieces into small individual crystals.

Etchant: 5 gm NaOH, 5 gm $K_3Fe(CN)_6$, 100 ml H_2O. $\times 213$.

Inclusions originating from a melting and pouring operation are grouped into two classes—those which are precipitated from the molten or freezing alloy and those which are entrapped by turbulence between metal and slag (or flux) or eroded from furnace and ladle walls and spouts. Usually the two classes can be differentiated on the basis of size and distribution. The melt reaction products are usually small and quite uniformly distributed, whereas the entrapped inclusions can be very large and almost always haphazard in their distribution. An unusually large such inclusion is shown in Fig. 1.26. The higher magnification view serves to reiterate, as elsewhere in this book, that ceramic materials in the crystallized form have microstructures similar to alloys.

(a)

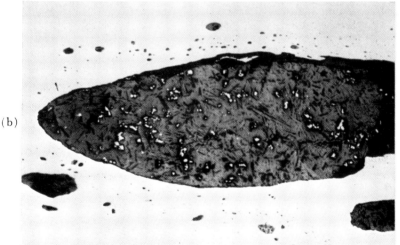

(b)

Fig. 1.26. These two companion micrographs show the existence (b) and structure (a) of an unusually large inclusion found in an ingot of AISI 4130 steel. This inclusion is particularly interesting because it has clearly undergone a crystallization history of the same nature as one can encounter in metallic materials. The structure shows two distinct primary crystallization products—the light primary phase is aluminous spinel, the dark primary phase is fayalite (Fe_2SiO_4)—and what appears to be a eutectic matrix of fayalite and glass. The actual identification of these phases requires other techniques such as X-ray diffraction or thin section petrography.

(a) Unetched. $\times 225$.
(b) Unetched. $\times 45$.

The appearance of inclusions generated by reactions in the melt and by the freezing operation depend very much on solubility conditions. There are inclusions which form by a deoxidation reaction producing a highly refractory oxide such as Al_2O_3 which is solid at all operating temperatures and insoluble in the melt. These will have no specific morphology. Nonmetallic phases which are soluble in the melt at high temperatures but separate into two liquids on cooling lead to globular shaped inclusions. The globules of nonmetallic liquid may crystallize into single or multiphase structures. The manganese-iron sulfide and certain sulfide-silicate inclusions in steel are of this

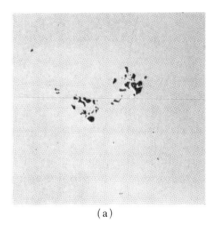

(a)

Fig. 1.27. These two micrographs were taken from castings of AISI 4130 steel. Both of the castings were partially deoxidized with silicon. Micrograph (a) represents inclusions in the steel with no further deoxidation. To the steel represented by (b), ferro-columbium and aluminum were added to the ladle in the amount of 4 lb per ton of metal. The inclusions in (a) are typical of the complex sulfide–silicate type. In (b), the proximity of the interdendritic columbium carbide–iron eutectic shows that the sulfide–silicate nonmetallic inclusions are swept into the last melt to freeze. Unetched. ×350.

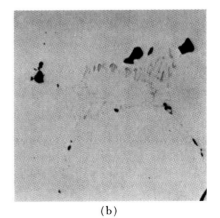

(b)

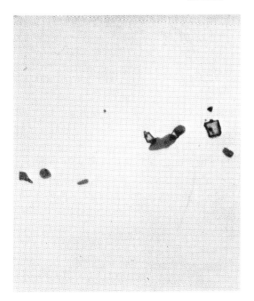

FIG. 1.28. The sharply angular, cubic shaped inclusion in the cast steel is titanium carbide. These inclusions are actually oxy-carbides, nitro-carbides, or oxy-nitro-carbides because titanium is a powerful scavenger of all three interstitial elements and the phases TiC, TiN, TiO are of like structure and completely miscible. In the unetched sample condition, these inclusions possess various shades of pink and gold. The sharply angular shape of these inclusions indicate that they crystallized from the melt before the steel began to freeze. In this case the crystallization was not necessarily the result of cooling. The titanium is added shortly before pouring. It dissolves into the steel and proceeds to combine with the ambient interstitial elements. The melt saturation for TiC is small and the rest crystallizes isothermally as the scavenging proceeds. The sulfide-silicate complex is apparently nucleated by TiC because these inclusions are nearly completely enveloped. The steel is AISI 4130 deoxidized with silicon and to which has been added as a ladle action, a commercial Ti-Al-B grain refiner in the amount of 4 lb per ton. Unetched. ×500.

type (Fig. 1.27).[*] Inclusions which crystallize from the melt as primary phases have well-defined angular shapes such as the titanium carbide in Fig. 1.28. Finally there is the inclusion species which derives as a secondary crystallization product from the melt and is disposed about dendrite or grain boudaries in network form, sometimes, at three grain intersections showing clearly a eutectic-type structure. Many inclusions are formed by chemical reaction of alloying additions and contained impurities. When the products of reaction are of high melting point, the residue in the cast structure will be in fine particulate form, usually randomly dispersed. When conditions permit settling out, the top of a casting may contain clusters of inclusions

[*] See S. L. Case and K. R. VanHorn, "Aluminum in Iron and Steel," pp. 52–77. Wiley, New York, 1953.

which have coagulated during gravity floating. The most finely divided inclusions come from deoxidation-type reactions which initiate as the metal cools and begins to solidify. These often distribute preferentially around the periphery of dendrites.

Hot working imposes a clear directionality to the distribution of nonmetallic, included materials. Glassy slags with low softening temperatures (originally in globular form) become stretched out into long thin strips by hot working as in Fig. 1.29. Clusters of more refractory oxides become strung out in a line, although the individual particles themselves are not deformed. Laps and seams from folded-in surface oxide films appear as continuous thin bands of oxide leading toward and sometimes right to the surface. That the oxidized seam sometimes ends beneath the surface is primarily significant of the

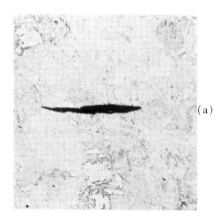

(a)

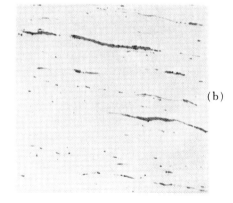

(b)

FIG. 1.29.   Micrograph (a) was taken from the interior of a thick-walled, tubular steel forging (AISI 4130). The large black inclusion is slag entrapped during pouring and solidification of the ingot. The original spheroidal form of the inclusion has been deformed to a lenticular shape by the forging operation. Other inclusions can be seen but the mixture of tempered martensite and bainite in the etched structure makes them quite difficult to resolve.

The second micrograph (b) shows the sharp resolution provided by the unetched condition. The specimen is taken from a forged gear blank of a resulfurized steel. The large manganese sulfide inclusions have been stretched out in the direction of principal strain. During the freezing process the manganese sulfide was enveloped by a skin of silicates. These are not as ductile as the sulfide, and as a result the envelope was ruptured repeatedly leading to the black fragments at the sulfide-steel interface.

(a) Etchant: 2% HNO₃, 98% ethyl alcohol. ×175.

(b) Unetched. ×70.

limitation of a two-dimensional section in characterizing a three-dimensional form.

Nonmetallic inclusions are best viewed in the unetched condition. Their natural colors, reflectivity, and surface roughness make them more clearly resolvable. Actually the etching exaggerates their size as well as confuses the viewer with the surrounding detail of structure.

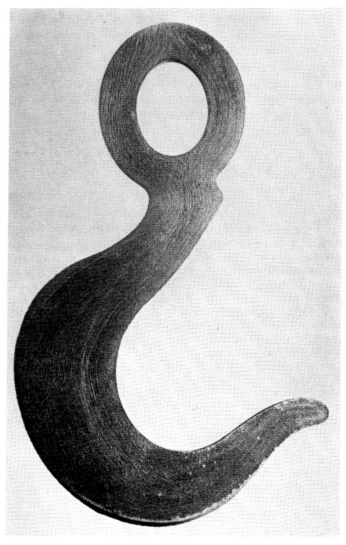

FIG. 1.30.

The etching rate around the periphery of an inclusion can be more rapid than elsewhere with the result that the inclusion appears larger than it really is. However, it is good practice to view inclusions in both the unetched and etched state and in that order. The unetched state permits appraisal of their real shape, size, and distribution. The etched state correlates the inclusions with various aspects of the metal structure. The intergranular oxide distribution of certain internally oxidized conditions has to be appreciated in the etched state.

It is never redundant to repeat that much can be learned from both low and high magnifications—each brings out different but related observations. The macroetching of a forging is a case in point; see Fig. 1.30. By heavily overetching a ground section, the strung out inclusions are brought out in sharp relief. The result is a fibrous pattern easily visible to the naked eye, revealing the principal directions of deformation during the prior forging history. This permits also interpretation of the directions of minimum ductility and toughness which are the directions, at any point, transverse to the lines of extension of the fibrous inclusions.

In the alloys of aluminum, considerable mechanical property anisotropy can exist without apparent inclusions in the microview. The fiber pattern, however, clearly develops from macroetching. In this case the fibered appearance derives from banding. The dendritic

---

FIG. 1.30. In the original cast ingot, much of the inclusion population is concentrated in discontinuous films around the periphery of each dendrite. The dendrites themselves possess a segregation pattern. With heavy deformation, the films of inclusions become strung out and the segregation patterns distorted according to the principal directions of shape extension. Neither effect is affected by recrystallization and transformation processes. They remain as permanent record of the principal directions of deformation. Their presence is brought out by deep etching in aggressive acids producing a relief effect. The fiber pattern visible to the unaided eye is a combination of the inclusion distribution and banding (see Fig. 1.22).

The macrograph is of a forged steel crane hook. The fiber pattern shows the hook was formed from bar stock rather than blanked from a heavy plate. This is indicated by the conformity of the fiber pattern to the hook profile except at the extremity of the eye of the hook. The emergence of the fiber pattern to the surface of the forging indicates that the eye itself was formed by a punching and expanding operation.

The fiber pattern indicates anisotropy of ductility, toughness, and fatigue strength. Typically, emergence of the fiber pattern to the surface produced by improper forging design or subsequent machining will correlate with preferred sites of crack initiation.

Etchant: 5 gm $FeCl_3$, 10% HCl, 10% HNO_3, 5% H_2SO_4, 75% H_2O. $\times 0.6$.

segregation in the original ingot is largely preserved although strung out in the wrought structure. A direction perpendicular to the long axis of deformation presents a profile of sharp changes in major or minor chemistry on a microscale. Differences in dissolution rate in acids yields a real profile of sharp grooves and ridges. Banding is probably most often the origin of fibered macrostructures.

CHAPTER II

Metallography of Fracture

Diagnosis of the origins of a failure usually requires several kinds of information. These relate to the material, the fracture appearance, and the circumstances associated with the cracking. Metallography provides the important input on details of fracture appearance, the microstructure through which the crack moved, and, by inference, something about the circumstances.

The history of a service crack usually involves three steps or stages. The crack nucleus can be some flaw that intrinsically has a crack root sharpness, a flaw that acquires the sharpness of a crack root early in load excursion history, or an inclusion of a material that is so brittle that it cracks on the first load excursion and continues to propagate into the surrounding matrix. Unless the crack nucleus is abnormally large or the material is of abnormally low toughness, there follows a period of slow crack growth. There are several mechanisms for slow crack growth, and it is usually of great importance in failure analysis to identify which mechanism or mechanisms were operative.

At some critical crack size, which is governed by the toughness of the material (or its tensile strength) and the operative net section stress, the crack growth will change abruptly from slow to very rapid. The crack appearance in the fast-propagation stage provides important clues as to the conditions of service and the correctness of choice of the material.

The metallography of fracture involves visual inspection, low-power stereoptic light microscopy, polished-section light-reflection microscopy, and the scanning electron microscope. Each of these provides some types of important information that the others cannot. Replica electron microscopy may be necessary to provide the information that light microscopy cannot because of the inadequacy of resolution. The interpretations of microstructural features are basically the same.

Visual inspection of the fracture surfaces followed by low-power stereoptic light microscopy provides an overview of the crack growth

49

progression. In many cases, markings (Fig. 2.1) point out the direction of crack motion so that it is possible to follow indicators back to the slow growth zone and even to the nucleus. Visual observation of the absence or presence (and amount) of shear lip and lateral contraction are important clues to proper diagnosis. Having located the subcritical, or slow-growth, crack zone, one can cut out specimens for examination at higher magnification both of the fracture surface and the path of the crack through the microstructure.

The progress of the crack through the microstructure is an important observation, particularly when the path follows some systematic feature of the crystals, distribution of phases, interfaces, or inclusions. For this aspect of the investigation to be effective, it is necessary to locate and extract a sample containing a crack root intact with both separated surfaces and the forward, uncracked material. In the case where a single crack has run to the extent of complete separation of fragments, this may not always be possible. But there are more often instances of multiple crack generation or crack branching where regions of arrested crack growth can be found.

The scanning electron microscope by its depth of focus is especially valuable in providing sharply resolved details of the fracture surface profile over a useful range of magnifications from ×50 to ×15000. The fine details of the fracture surface have appearances that are typical of specific fracture modes. The fracture modes, in turn, allow conclusions about the circumstances responsible for the failure.

Fractures are divided into two primary categories—ductile and brittle. For relatively small cross sections (or thicknesses), ductile fractures are associated with measurable lateral contraction, or "necking." Also, that part of the fracture path at and near the surface follows a shear or slant plane with respect to the direction of dominating tensile stress, while the fracture path in the interior follows a generally flat

FIG. 2.1. A crack may run a long distance from its point of origin. Markings on the fracture surface of the fast running crack may be used to point back to the origin. Figure 2.1(a) shows the so-called "chevron" markings on the fracture of a steel plate. The upper view shows the chevron markings some distance from the fast-crack initiation point. The lower view shows the origin and the transition in markings from radial marking to chevron markings. Figure 2.1(b) shows the detail of a pattern of radial marks pointing toward the fracture initiation site in the tensile fracture surface of a specimen of an epoxy plastic containing a dispersion of rubber particles.

(a) ×1.
(b) ×11.

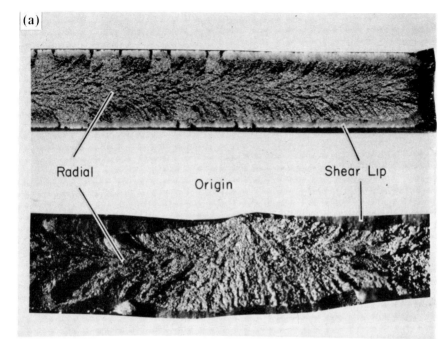

Radial   Origin   Shear Lip

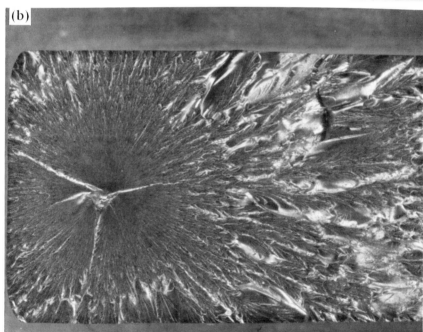

plane perpendicular to the main tensile stress. In a cylindrical tensile specimen, this is the familiar "cup-and-cone" configuration. For relatively thick cross sections of the same material, the fracture surface—although ductile—will be predominantly flat with little or no slant component at and near the surface. Thus, for a given material, fractures with increasing plate thickness or bar diameters progress from nearly all slant to nearly all flat. However, the fine structure of the fracture surface in both cases has a typical fibrous aspect that makes the identification of the ductile mode quite positive (Fig. 2.2).

Brittle fractures are flat and transverse to the axes of major tensile stress. The flat fracture extends from surface to surface, and there is no measurable lateral contraction. Fracture surface features are divided into two distinguishable categories—transcrystalline and intercrystalline. The term "cleavage" is often appended to each of these brittle fracture appearances. This implies that even at the microstructural level of examination there is very little evidence of plastic deformation near the crack surfaces and ahead of an arrested crack root.

Intercrystalline cleavage has the most recognizable morphology, because it reveals the faceted geometry of the individual crystals in the assembly (Fig. 2.3). Intercrystalline cleavage is also identifiable with specific circumstances of the failure, such as immersion in a stress–corrosion environment, contact with a liquid metal (wetted to the surface) (Fig. 2.4), or hot shortness (Fig. 2.5). Intercrystalline cleavage can also be indicative of the selective distribution of a brittle phase at grain interfaces or an embrittling segregation. However, this must be verified. Cross-section light optical metallography can demonstrate the presence of an intercrystalline film or network of a brittle phase and its conformity to a crack path. The distinction between intercrystalline and transcrystalline cracking in certain circumstances can permit discrimination between different possible stress–corrosion factors.

FIG. 2.2. Scanning electron microscope views at two magnifications—(a) ×1000 and (b) ×10,000—illustrate the definitive appearance of a ductile fracture surface. The ductile, or fibrous, fracture has a tufted appearance that results from myriad local initiations of necking, which draws down during fracture extension to points and edges. The concept of the mechanism of ductile fracture invokes the generation of voids in the plastically deforming material. Material lying between adjacent voids have a "necked" geometry, which makes for void growth and material thinning during the fracture extension regime.

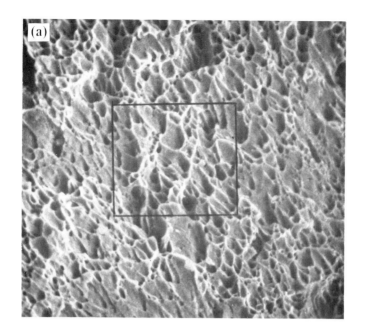

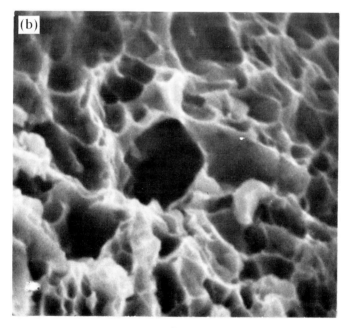

53

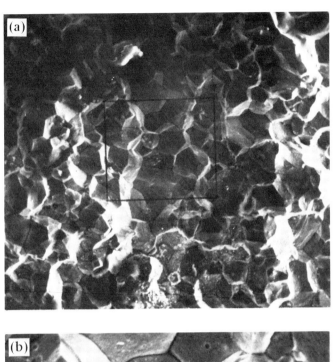

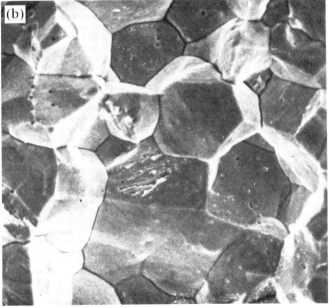

Transcrystalline cleavage is less easily recognizable in all of its forms. Where the cleavage represents separation along crystallographic planes and directions in relatively large crystals, the high magnification "river" pattern is characteristic (Fig. 2.6). This type of cleavage also has a specular appearance in visible light. When the crystals are very small, as with tempered martensitic steels, the fracture morphology is rather featureless. Lack of features is true of the brittle fracture of face-centered cubic solid solutions. These featureless transcrystalline fracture surfaces usually allow little definitive conclusion except as a default of other more typical appearances.

It should be noted at this point that mixed modes of fracture are very common. A mixed mode is not the same as sequential modes of fracture. Thus, for example, in carbon steels fractured in the ductile–brittle transition temperature range, the fracture topography from ferrite crystal to crystal may alternate between fibrous and transcrystalline cleavage. In stress–corrosion cracking, the intercrystalline cleavage fronts may isolate ligaments that fail ultimately in a fibrous mode. These are both instances of mixed modes of fracture. On the other hand, a cracked inclusion (brittle and cleavage) can initiate fatigue cracking (flat–featureless or flat–striated), which upon reaching a critical crack dimension will convert to fast ductile (fibrous–flat and slant mixed) fracture, which leads to failure. This represents a common sequence of changing cracking modes.

Crack Nucleation

Slow cracks most often initiate at or near the surface, partly because bending components of the stress system produce maximum tensile stresses at the surface, and partly because macrostress concentrations such as sharp cornered shoulders, grooves, slots, and even deep machining marks exist at the surface. These circumstances conspire with the chance occurrence of material flaws of cracklike character existing near the surface and particularly near a stress concentration (Fig. 2.7).

FIG. 2.3. Scanning electron microscope view of intercrystalline cleavage at (a) ×300 and (b) ×1000. It reveals the polyhedral form of the crystals in a polycrystalline aggregate that fills space. This pattern of fracture is associated with hot shortness, certain types of stress corrosion cracking, and grain boundry networks of brittle phase. Other observations, such as evidence of oxidation or corrosion, circumstances of failure, and optical reflection micrography, are necessary to identify the nature of the fracture mechanism.

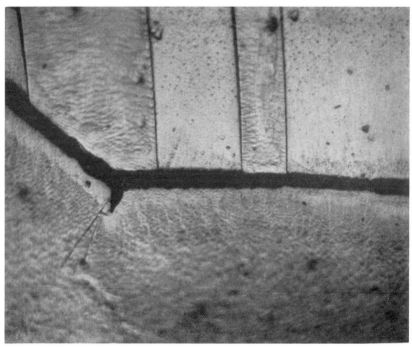

FIG. 2.4. A 70/30 brass wetted with mercury will fail under tensile loading at a stress slightly above its yield point and after only a few per cent plastic elongation, whereas under normal conditions the annealed metal is capable of 30–50% extension before failure. Wetting with mercury is a classic example of embrittlement by liquid metals. In single phase, recrystallized metals such as the specimen of brass illustrated, the fracture path is invariably intergranular. This micrograph also provides some special information. Attention is drawn to the boundaries of the annealing twins in the large upper grain. Because the twin orientation is in a deep energy trough, twin boundaries intersect normal grain boundaries even at 90° angles and maintain this as a stable intersection. The twin boundary as seen under the microscope is always a highly perfect straight line. This linear perfection can be used as a gauge mark to measure approximately the amount of plastic deformation associated with the propagation of the crack. If some measurable plastic deformation did precede the advance of the crack, the twin boundaries should be visibly bent as if the grain boundary were a plane of shear. The twin boundaries are straight right to the fracture surfaces which were once matched to form a grain boundary. There is some small distortion apparent in one instance but this is because it is so difficult to preserve an unrounded edge in metallographic preparation. Most brittle fractures have easily resolved plastic distortion around the crack zone and it demonstrates the magnitude of embrittlement produced by wetting with mercury.

Etchant: 5 gm ammonium persulfate, 100 ml H_2O. ×450.

Fig. 2.5. A butt fusion weld constitutes one of the most severe tests of the hot tearing tendencies of an alloy. During the welding operation some of the heat generated is absorbed by the butt ends of the plates being joined. These ends expand and later while the weld pool is freezing they contract. In this freezing period (the joint is essentially rigid), crystallization proceeds from both butt ends and the base plates are contracting. Without design or process modifications the center plane of the weld pool is both the last to freeze and the recipient of the maximum contraction stresses.

Cracks resulting from such hot tearing are usually intercrystalline because this is the general pattern of distribution of residual melt. The micrograph shows a hot tear revealed by sectioning the weld zone. The weld represents a so-called bead-on-plate specimen in which a nonfusing welding electrode is used to fuse the butt edges of the specimen to provide the molten weld pool. The material is the nickel-based Inconel alloy which is a highly alloyed solid solution with nominal composition of 84% Ni, 14% Cr, and 6% Fe. The hot tear crack follows the plane between two large grains whose major orientation axes are inclined only about 20° from each other as indicated by the long axes of the coring patterns in each.

The orientations of the grains as indicated by their coring patterns also indicate the direction of heat flow during freezing. This weld was made by several passes of the torch each laying down a new layer of molten metal. This micrograph is taken in a region where the mass of the prior weld bead is a heat sink for the next. Successive weld beads therefore show a mixed pattern of grain growth during crystallization, depending on the dominance of chilling by either the butt ends of the plates or the already existing underlay of weld metal. (Micrograph courtesy of P. J. Rieppel, Battelle Memorial Institute, Columbus, Ohio.)

Etchant: Aqua regia. ×100.

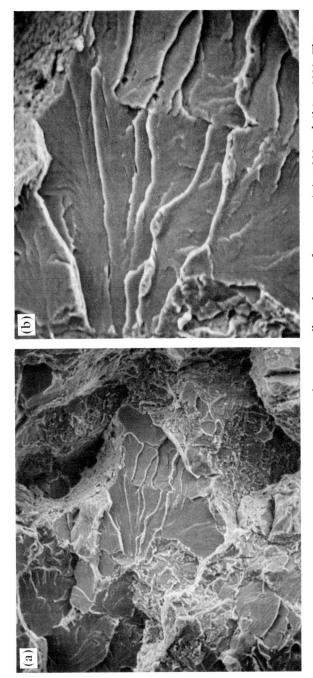

FIG. 2.6. Scanning electron microscope appearance of transcrystalline cleavage fracture at (a) ×300 and (b) ×1000. This type of fracture is characterized by very flat plateaus separated by short steps. The plateaus are associated with a crystallographic plane identity. The steps follow a wavy direction approximately parallel to the direction of fracture propagation. The family of steps gives the appearance called "river pattern." The "river pattern" is confined to one crystal. The fracture in adjacent crystals may have similar but, always in some manner, a different pattern. Individual crystals in an aggregate can show either cleavage or fibrous fracture appearances.

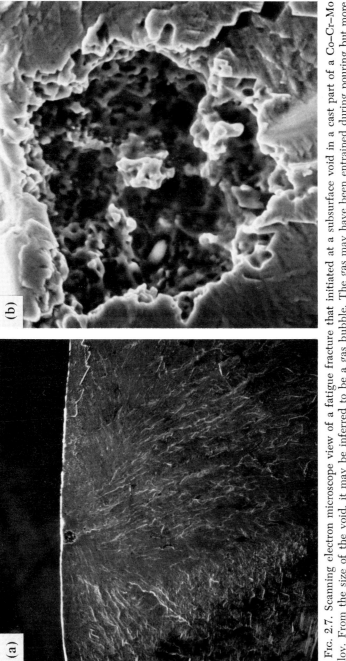

Fig. 2.7. Scanning electron microscope view of a fatigue fracture that initiated at a subsurface void in a cast part of a Co–Cr–Mo alloy. From the size of the void, it may be inferred to be a gas bubble. The gas may have been entrained during pouring but more likely was expelled from liquid solution during freezing. Although spheroidal in shape, this void acts as a stress concentrator on the thin bridge of metal separating it from the surface. A closer view of the interior of the void shows the intrusion of growing dendrites that provides the multiple notch and stress concentration effect. (a) ×95. (b) ×1900.

It is relatively uncommon for the scanning electron microscope or for cross-section light reflection microscopy to identify the actual original flaw located at the edge of or in the middle of the slow crack growth zone if it is an inclusion. However, very large nonmetallic inclusions, large crystals of intermetallic compounds, shrinkage and gas porosity, seams, or laps (Fig. 2.8) can be readily recognized as the crack initiator. Often it is sufficient to recognize the presence of any of these near the fracture surface to deduce that another of like dimensions was probably responsible for the crack itself.

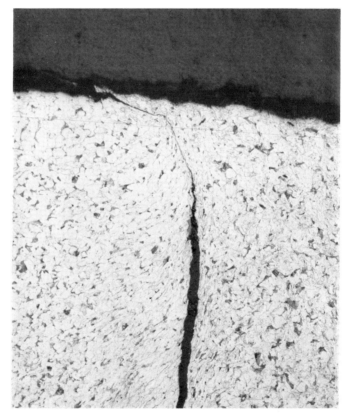

Fig. 2.8. A seam in hot-rolled carbon steel rod escaped detection at the mill. The batch of rod was used in the manufacture of bolts, which requires a heading operation. The large plastic strains associated with heading caused the seam to open up and become readily visible. Subsequent inspection of unused rod revealed the seam. The micrograph shows that near the surface the seam was nearly welded shut. However, the reality of separation is still evident.
Etchant: nitol. ×100.

Strings of inclusions or voids (Fig. 2.9) should be regarded as a single flaw of the length dimensions of the whole string for purposes of judging stress concentration or crack nucleating severity. This is because the ligaments of metal separating the particles or voids can fail individually at low stresses. This would not happen in the matrix when the microflaws are more widely and uniformly distributed. When the ligaments fail, the microflaws link up to form macroflaws.

Crack branching can be stimulated by recognizable microstructural conditions. The so-called "laminating" crack is usually a branch or series of branches that propagate in planes perpendicular to the main

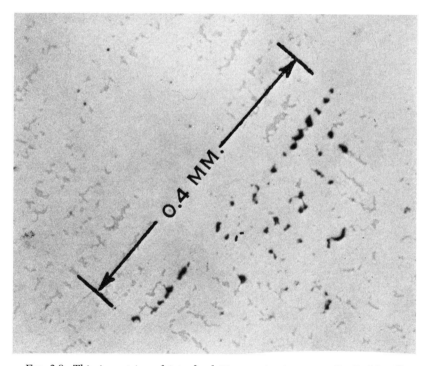

FIG. 2.9. This is a string of interdendritic porosity in a cast Co–Cr–Mo alloy. They are probably shrinkage voids, although a small amount of soluble gas precipitation could produce a similar appearance. However, with gas porosity one would expect to see some larger voids as well in a nondendritic configuration. From a service viewpoint one must expect that the metal ligaments between the voids would fracture early in the fluctuating stress experience so that the string of voids would become a cracklike flaw of the dimensions shown. By examining, as in this case, in the unetched condition, the maximum contrast is attained between the interdendritic voids and the interdendritic carbides, which in this micrograph are barely visible. Etching darkens the carbides so that the voids become indistinguishable.

Fig. 2.10. The composite picture of (a) results from bending and then back-bending a narrow strip (0.020 inch thick) of carbon steel. A crack formed in the plane of the strip and by propagation and back-bending literally divided the thickness into two halves. Hence the term "laminating crack" is applied. The microstructure of the unfailed material adjacent (b) shows the origin as a continuous oxide-filled seam that probably represents rolled "pipe" residual from an improperly cropped ingot. (a) ×30. (b) ×200.

or original crack. This condition can be caused by a high density of ribbonlike inclusions lying in the planes of lamination (Fig. 2.10). They can be recognized metallographically. Sometimes the ribbonlike inclusions themselves cannot be resolved, but their presence is indicated by the ribbon shape of recrystallized grains. Laminating tendencies can be provoked by highly developed textures wherein a cleavage plane is lying in the plane of the sheet. This condition is not revealed by metallographic techniques.

Stress–Corrosion and Other Environmentally Induced Cracking

A number of aqueous and gaseous environments induce brittle cracking in specific materials. For example, ammonia vapor and ammoniacal or amine solutions are specific to copper-base alloys. On the other hand, saline solutions are operative on a number of diverse materials— aluminum alloys, austenitic stainless steels, and high-strength martensitic steels. In each case of an aggressive environment operative on a specific material, the crack path is specific—either intercrystalline or transcrystalline. However, a high order of specificity is sometimes required. For example, modifications in complex ammonia ion concentration and species as well as pH may change the crack pattern of brass from intercrystalline to transcrystalline.

Caution must be exercised in arriving at a firm conclusion that stress–corrosion mechanisms are operative. In aluminum alloys, simple intergranular corrosion can produce cracklike flaws that are indistinguishable from stress–corrosion cracks. In fact, the two processes can proceed concurrently and can be distinguished only with difficulty (Fig. 5.26). On the other hand, the transcrystalline stress–corrosion cracking of austenitic stainless steels is readily distinguished from intergranular corrosion.

Stress–corrosion cracking initiates at the surface, generally at no resolvable nucleation site. Current theories suggest that active slip bands can be nuclei. Accordingly, it is not surprising that many cracks nucleate simultaneously. This is a point of distinction, because other brittle cracking processes are not usually associated with a high multiplicity of surface cracks. Of course, if a significant stress concentration feature exists at the surface, a single stress–corrosion crack will generate at the point of highest local stress, and the propagation of that crack will relax stresses in the immediate vicinity.

Stress–corrosion cracking at high stresses is often characterized by a high frequency of branching (Fig. 2.11). Thus, cross-section light

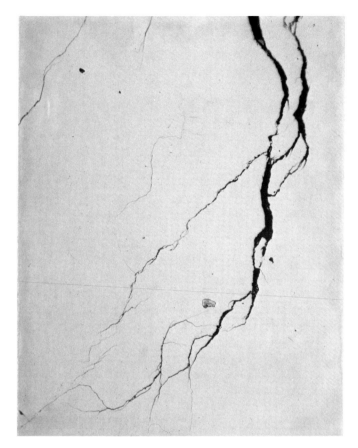

FIG. 2.11. Hot chloride solutions produce in stressed austenitic stainless steels both multiple cracks and multibranched cracks. Many of the crack branches are hairlike and only resolvable in the unetched condition. The fact that some hairlike cracks seem isolated from larger cracks serves to emphasize the planar limitations of metallography in describing a spatial complex cracking arrangement. This multibranch configuration combined with the transcrystalline path revealed by etching certainly identifies the stress–corrosion origin of the failure for this particular material.
×100.

reflection microscopy will show many sessile secondary cracks emanating from the primary crack, and the crack tip will show a multibranched arrangement of probing cracks. Presumably, in such a configuration one branch runs ahead of the others and the remainder become sessile. Branches in anisotropic materials travel great distances causing an exfoliated fracture that is really many parallel laminating cracks. Exfoliation can also begin at the edges of plates of susceptible alloys even though the stresses through the thickness direction are relatively small compared to those in the plane of the plate.

Wetting by certain liquid metals can cause cracking in certain stressed engineering alloys. As with stress–corrosion cracking there is a specificity in the combinations of structural solid and the environment. Thus, for example, liquid mercury is specific to aluminum and copper alloys but is ineffective on iron alloys and steels. Cracking initiates at the liquid–solid interface without evidence of crack nuclei. However, there is rarely more than one crack nucleated, presumably because the first crack grows at a much faster rate than stress–corrosion cracks. As with stress–corrosion cracking, the crack will always begin preferentially at a stress concentration, even if it is only a machining mark.

Liquid metal cracking is predominantly intercrystalline. Such cracks can propagate with remarkably little associated plastic deformation and in that sense is one of the most brittle processes (Fig. 2.4). Transcrystalline cracking can be seen by cross-section light microscopy only in the cases of cold-worked metals or highly elongated recrystallized grains (grain growth anisotropy) when the principal tensile stress is planar with the ribbon configuration of the grains.

Hot shortness involves incipient melting or residual liquid from freezing existant at the time tensile stress is applied (Fig. 2.5). This can produce the same effect as surface wetting with resultant intercrystalline fracture. An example is leaded martensitic steels. The condition of fracture and the fracture appearance are the same whether the lead is distributed internally as separate particles or wetted to the surface of an otherwise unleaded steel of similar alloy and carbon contents. Probably all forms of hot shortness are a manifestation of liquid metal embrittlement despite the absence of a resolvable distribution of low-melting liquid. Bismuth embrittlement of copper is a case in point. Very small additions of bismuth to copper during melting will produce severe hot shortness, although the bismuth is not resolvable in the cast structure. Wetting the surface of pure copper with liquid bismuth will produce an identical hot short fracture under stresses of yielding or plastic deformation.

Hydrogen embrittlement or hydrogen cracking is a form of environmentally induced brittle fracture. Generally, it is supposed that the environment contains or produces hydrogen that dissolves into the metal, and the cracking process begins in the region of an appropriate combination of high stress and high hydrogen content. The crack morphology can be transcrystalline or mixed trans- and intercrystalline. In this way it is not specific. It can also produce laminating cracks, since a major concept of mechanism involves precipitation of pressure-generating molecular hydrogen at metal–inclusion interfaces. Thus, if the inclusions have a ribbonlike shape, the cracks can lie in a plane parallel to the major surfaces.

A special form of hydrogen embrittlement is the flaking of castings and heavy forgings. Internal cracking can result from a hydrogen crack forming a nucleus, which grows by fatigue to critical size and then transforms into a catastrophic fast crack. In such a case, the sequence can be readily appreciated visually on the fracture surface by the appearance of a flat, featureless nucleus, a concentric zone that is less featureless, and a large outlying fracture that is fibrous.

Fatigue Cracking

Fatigue is a form of slow and brittle crack growth associated with fluctuating tensile, tensile–compressive, or shear stresses. It can occur in any metallic material regardless of intrinsic ductility as exhibited by a conventional tensile test. Fatigue cracks grow at measurable rates under stress limits of fluctuation that are much less than the yield stress. So in the region of the fatigue growth of a crack, there are few evidences of general yielding. However, if the surface is polished to a high finish or if conditions are appropriate to aging decoration, slip bands can be seen at and near either surface of a crack (Fig. 2.12) and in advance of a crack root. In some cases, the earliest stages of fatigue cracking follow slip band planes, from which it has been deduced that slip activity is fundamental to some nucleation of fatigue cracking. Since it has been clearly shown that the endurance limit of steels is related to inclusion population, one may conclude that cracked inclusions are a more common basis for fatigue nucleation.

FIG. 2.12. Scanning electron microscope views of stage I fatigue cracking [(a) ×930] and of the transition from stage I to stage II [(b) ×1100] in a cast Co–Cr–Mo alloy. The part during manufacture had to be polished to a very good finish. Service stresses generated slip bands, which are easily visible in (a). Fluctuating stresses ultimately nucleated fatigue cracks in certain of these slip bands as can be clearly seen also in (a). The fracture surface (b) shows the transition from the flat featureless appearance of stage I crack growth to the striated appearance of stage II crack growth.

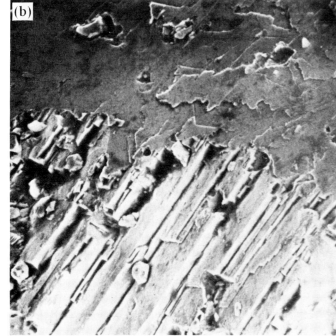

FIG. 2.13. This is a composite of scanning electron micrographs that present a large field of view of a fatigue fracture surface. The inset (b) shows the typical striated pattern. However, the large field (a) shows that only a small part of the whole fracture surface demonstrates this fracture profile. Most of the fracture surface is essentially featureless. Therefore, it must be recognized that the absence of the striated pattern cannot be taken as evidence that the fracture mode is other than fatigue. On the other hand, the appearance of the striated pattern can be taken as positive evidence that fatigue is involved. Notice that crack growth rates have been estimated.

(a) ×200. (b) ×1000.

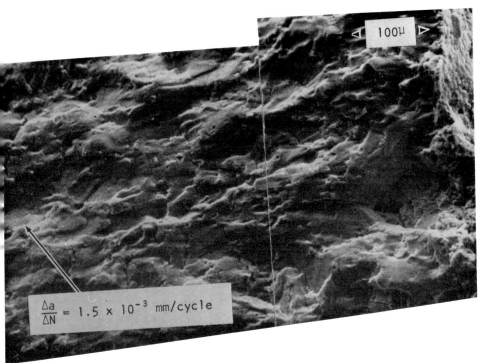

$$\frac{\Delta a}{\Delta N} = 1.5 \times 10^{-3} \text{ mm/cycle}$$

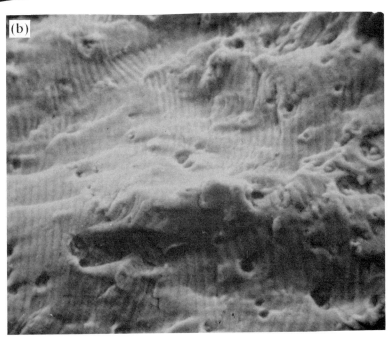

(b)

In fine-grained metals, the majority of the fatigue crack growth follows a plane perpendicular to the largest tensile stress. Often but not always this fracture surface shows a characteristic series of concentric or parallel striations (sometimes called "beach" marks) (Fig. 2.14). The distance between striations seems to represent the incremental advance of the crack for each stress fluctuation. The striated pattern can sometimes be seen with only ×30 magnification in a light stereoptic microscope but more often requires magnifications of ×500 to ×10,000 with the scanning electron microscope.

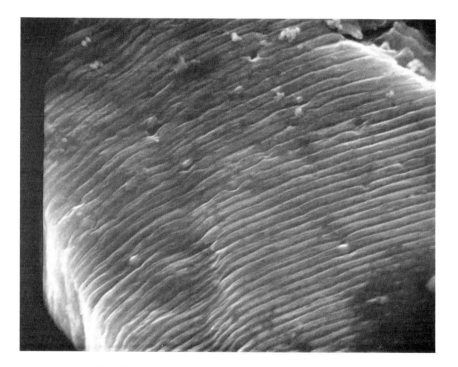

Fig. 2.14. The "beach marks" or "striated" appearance is unique to the appearance of fatigue fracture surfaces. This exhibit is taken by scanning electron microscope from the fracture surface of an aluminum alloy (6070-T6) pressure vessel that failed by repeated pressurizing and depressurizing. If it is assumed that the distance between each striation corresponds to the crack advance of one cycle, the rate of crack growth can be easily computed from the magnification.
×6000.

Most often, the striated pattern can be seen only in isolated portions of the fracture surface (Fig. 2.13). Between these isolated portions, the fracture surface is highly variable in appearance and not really amenable to useful description. This featureless character is certainly not simply flat. There are all sorts of what appear to be torn ligaments, yet the surface is far from the classic fibrous appearance. Some fracture surfaces show none of the striated regions. From a diagnostic viewpoint, the evidence of striations can be taken as a positive indication of fatigue cracking, but its absence cannot be used to argue against fatigue as the mechanism of slow crack growth.

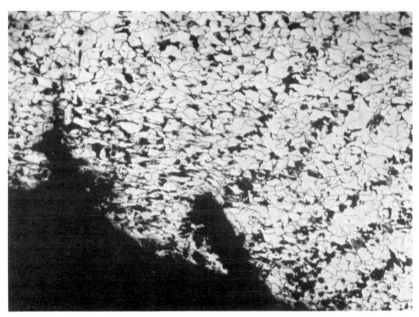

Fig. 2.15. The energy of fracture involves not only the work of forming new surfaces but also the plastic energy dissipation in regions immediately adjacent to the fracture surfaces. The existence of the plastically deformed zone is illustrated in this micrograph taken at the root of the V-notch in a Charpy impact specimen which has been struck with sufficient force to start a crack. The material itself is a low-carbon steel. The banded distribution of pearlite (black regions) indicates a retained segregation inherited from the ingot condition.
Etchant: 2% HNO$_3$, 98% ethyl alcohol. ×120.

Fatigue cracking is usually transcrystalline. There are rarely evidences of sessile cracks at the surface near initiation of the primary crack. The advancing crack itself will be seen often to fork, with one prong becoming sessile, but fatigue cracks do not multibranch.

Toughness Indications

Low or high toughness is a condition that can be recognized using cross-section microscopy by evidence of local plastic deformation near the crack (Fig. 2.15) and by the crack yawning or crack root sharpness (Fig. 2.16).

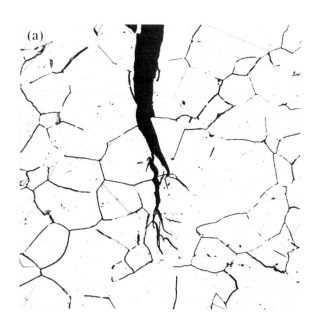

FIG. 2.16. Austenitic stainless steels under stress corrosion crack in boiling, saturated MgCl₂ solution. The cracks are sharp and transcrystalline (a and b). If the fluid environment is removed and the stress increased, the material will regain its high toughness, the cracks will yawn, their root radii will increase, and they will not increase in depth. The grains in the vicinity (c) will show evidence of large plastic strains.

Etchant: 5 gm CuCl₂, 10% HCl, 90% ethyl alcohol. ×175.

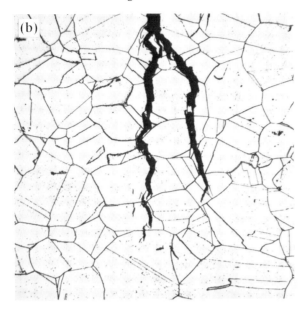

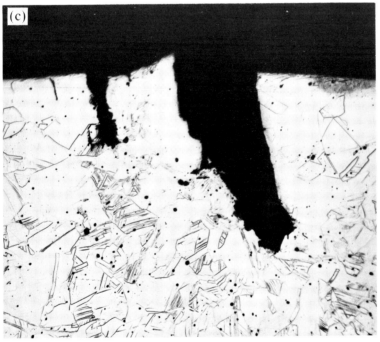

Creep Rupture

Rupture under static load at elevated temperatures is the culmination of nucleation of many intergranular separations and their final link-up as a continuous long crack. Apart from the most commonly intergranular character of creep ruptures, the appearance of many local grain separations near but not connected to the crack is typical (Fig. 2.17).

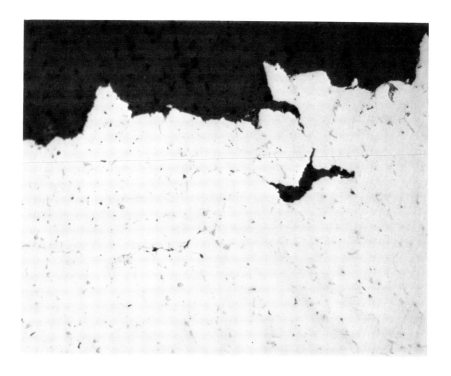

Fig. 2.17. There are applications of metals involving elevated temperatures and substantially static loads. The cast Co–Cr alloys are frequently used for their resistance to creep and creep–rupture. This specimen shows an interdendritic separation removed from but near the creep–rupture surface. The fracture surface itself represents a linkup of many of these separations.
×200.

CHAPTER III

Crystallization

Dendritic Crystallization

When the temperature at any point in the melt drops below the liquidus temperature, nucleation of solid crystals initiates. Most commonly, these crystal nuclei are of one phase species and we speak of a primary crystallization phase. In simple alloys there may only be one crystallizing phase. In more complicated alloys there may initiate at critical lower temperatures, secondary and tertiary crystallization implying in each case the nucleation of second and third phases. At singularities of composition and temperature, concurrent nucleation of more than one phase can occur.

The growth of nucleated crystals is governed by alloy concentration gradients in the surrounding liquid, by temperature gradients and the configuration of the surfaces of heat abstraction, and by the interference of other growing crystals. Apart from a few unusual cases such as the crystallization of graphite in nodular cast iron shown in Fig. 3.1, growth of crystals is highly anisotropic. This signifies that instead of growing as spheres, crystals grow as needles, plates, or even prismoidal shapes (see Figs. 3.2–3.4).

Crystals grow most rapidly in certain crystallographic directions. These preferred directions are characteristic of the crystal structure. Thus, face-centered and body-centered cubic metals and alloys grow most rapidly in [100] directions. The ⟨100⟩ family consists of three mutually perpendicular axes, so that a crystal can in principle grow as a six-pointed star shape. However, the planar arrangement of heat abstraction surface and the direction of thermal gradient perpendicular to this surface dictates that one crystallographic direction most nearly aligned to the gradient direction will dominate the growth options. Accordingly, to a first approximation, the crystal will grow as a needle and multiple nucleation will create a forest of needles oriented in the direction of the thermal gradient.

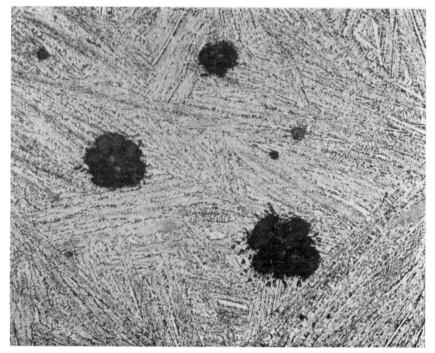

FIG. 3.1. This micrograph is intended to illustrate that nodular graphite in cast iron can be a primary crystallization product. A heat of cast iron inoculated with magnesium was shotted into water from successively lower melt temperatures. At some critical melt temperature, metallographic examination of the cast iron shot revealed a few nodules of graphite in a very fine matrix of graphite and white iron eutectic (carbide and transformed austenite). With successively lower melt temperatures, the number of observable graphite nodules increased. The melt temperature at which the cast iron shot first revealed graphite nodules can be taken as a close approximation of the liquidus for that composition. Shotting from above the liquidus causes such complete undercooling of the melt that primary crystallization is completely suppressed. (Cast iron with nominal composition: 3.62% C, 2.61% Si, 0.74% Mn, 0.10% S, 0.12% P.)

The larger graphite nodules show a dendritic structure while the smaller ones are almost perfectly spherical. While nodular graphite crystallization in cast iron can only be induced by the addition of special melt conditions or inoculants, rapid solidification rates alone are sufficient in Ni-C, Co-C, and Re-C alloys [see J. E. Hughes, *J. Less-Common Metals* 1, 377–381 (1959)].

Etchant: 2% HNO₃, 98% ethyl alcohol. ×225.

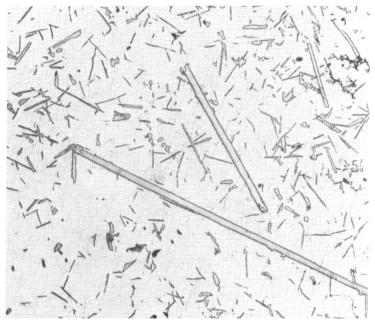

FIG. 3.2. Cast structure of an aluminum alloy containing 6% Fe. The long needles (or plates—one cannot make the distinction from this micrograph) are primary crystals of the $FeAl_3$ phase imbedded in a two-phase matrix which is a nonlamellar eutectic of aluminum solid solution (almost pure Al) and the $FeAl_3$ phase. In this type of structure one can distinguish the primary crystallization of $FeAl_3$ from the secondary crystallization product only by the relative size of particles.

Etchant: 1% HF, 99% H_2O. ×150.

Between the aligned crystal needles, secondary temperature gradients and melt segregation patterns are generated that stimulate nucleation and growth of lateral projections of the same crystallographic identity. Therefore, the needle configuration usually converts almost immediately to the branched or dendritic arrangement shown in Fig. 3.5. In the course of filling space by the transformation of liquid to solid, the branches will produce branches by a succession of nucleation events, and the branches thicken. Without feeding from remote positions, the volume change in the transformation will show up as microvoids between dendritic branches (Fig. 2.9).

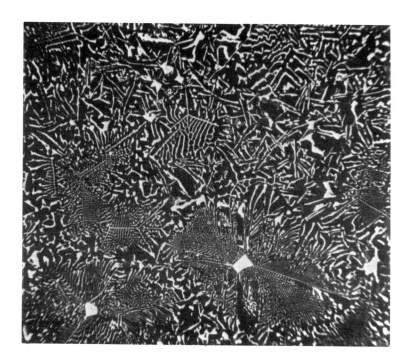

Fig. 3.3. Primary crystallization of Si in an Al-Si alloy containing about 25 w/o Si. These plate-like particles have sharply angular corners indicating that a relatively high anisotropy of interfacial energy existed between the growth nuclei and the eutectic melt. The eutectic itself is of the nonlamellar type, and there is clearly a strong tendency for eutectic divorcement around the primary Si crystals; that is, the Si of the eutectic prefers to grow on to the existing Si primary crystals leaving a band or region of Al solid solution in the near vicinity. Unetched. ×100.

Fig. 3.4. As-cast structure of a lead alloy containing 13.5% Sb and 6% Sn. This is a complex structure containing three phases—lead-base solid solution (etched black), intermetallic compound Sb-Sn (etched white), and antimony-base solid solution (minority gray constituent). The structure developed by the primary crystallization of almost cubic shapes of Sb-Sn around which, and nucleated by it, grew colonies of binary eutectic of Pb and Sb-Sn arranged in a "Chinese script" fashion with long dendritic spines of Sb-Sn. Finally crystallization was completed by the formation of a ternary eutectic of Pb, Sb, and Sb-Sn, the ternary eutectic being of the nonlamellar type.

Etchant: 15% HNO_3, 15% acetic acid, 70% glycerin. ×250.

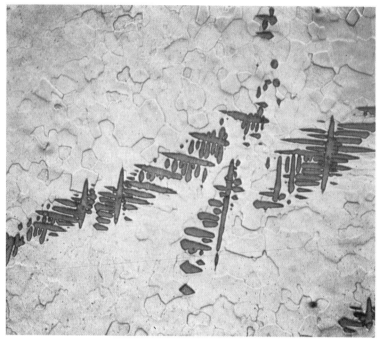

Fig. 3.5. A tin-base alloy containing 5% Zr in its as-cast form. This system has a steeply rising liquidus boundary at the Sn end, and a high superheat above the solidus temperature is necessary to get this amount of Zr into liquid solution. The large temperature difference between the liquidus and solidus permits the growth of large, branching dendrites of the primary crystallization phase, $ZrSn_2$. The orthogonal relationships between the branches of the dendrites indicate the low Miller indice character of the preferential growth direction. The matrix is a polycrystalline aggregate of tin grains. The tin phase has only an extremely small solid solubility for Zr, so evidence of precipitation is unlikely.
Etchant: 20% HF, 20% HNO_3, 60% glycerin. ×150.

If crystallization involves only one phase, the dendritic growth fills space and the dendritic pattern is only visible by etching procedures that emphasize composition segregation (see "coring"). But when primary crystallization is followed by secondary crystallization, then the primary dendrites are brought out in sharp contrast. Figure 3.6 illustrates this. What is seen is a main stem with second and third stages of branching. Because the dendrite is a three-dimensional shape, the polished cross section brings out only the traces of the shape. Thus,

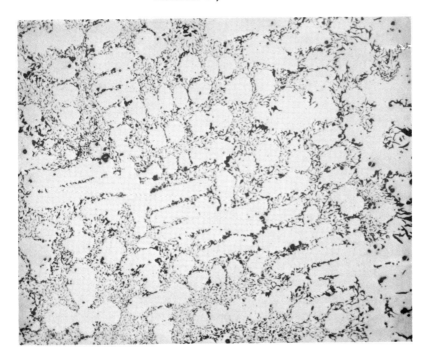

FIG. 3.6. Microstructure of a hypoeutectic gray cast iron (3.2% C; 2.1% Si; 0.4% Ti) showing the trace outlines of austenite (transformed to ferrite and pearlite) dendrites. The dendrites are outlined by a very fine (type D) eutectic of graphite and transformed austenite. The main, or primary, branch of the dendrite lies in the principle direction of heat flow.

Unetched. ×100.

the traces of the primary and secondary extensions are rodlike, while the tertiary branches are a series of closely spaced ellipses.

When primary nucleation is widely spaced, the secondary branches can achieve planar growth dimensions comparable to the primary branch. Thus, the individual crystals in the large cast grains shown in Fig. 3.7 are each only a single dendrite, but the growth and branching have been so unrestrained that the morphology of the dendrite in the late stages of solidification represented a lattice of thick plates something like an egg-crate arrangement.

FIG. 3.7. As-cast microstructure of a Co–Cr-Mo alloy. The investment casting process involved rather slow freezing in a hot ceramic mold, with the result that the cross section was occupied by a relatively few large, almost equiaxed crystals. Each crystal is a single dendrite, but the complexity of growth has erased the primacy of any one branch. The multiple branching of crystal growth has created a three-dimensional lattice of interleaving plates (parallel and orthogonal), which gradually thickened to fill space. The outline of the grid is produced by the "coring" effect.

Electrolytically etched in 10% ammonium persulfate solution in water. ×50.

Eutectic Crystallization

Simultaneous crystallization of two or more solid phases derives from a double (or multiple) supersaturation of the melt. The mutually influenced nucleation and growth product is called a eutectic. The eutectic composition in a binary phase diagram is at the junction of two descending liquidus curves. In a binary system, the eutectic crystallization occurs at or just below a fixed temperature. In polycomponent alloy systems, two-phase eutectic crystallization can occur over a range of temperatures.

The lamellar structure shown in Fig. 3.8 is usually given as the archetype of the eutectic crystallization structure. It is, in fact, only one of several structures developed by eutectic freezing systems. The lamellar eutectic appears as an assembly of colonies in each of which is disposed a discontinuous phase in the form of thin strips approximately parallel to each other. Within each colony, one phase is

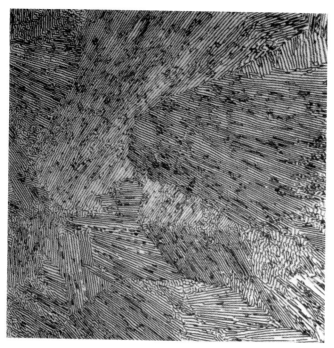

Fig. 3.8. This is the microstructure of Mg–45% Th alloy representing a well-developed lamellar eutectic of Mg solid solution and Mg₅Th. The growth colonies can be easily distinguished. Close inspection shows that the orientation relationships are regular because bands of Mg from one colony can be seen to be continuous with bands in another despite the change in direction. In this binary alloy system there is a strong tendency toward segregated eutectic structure. The present eutectic form was developed only after very slow freezing in the furnace.

Etchant: 1% HNO₃, 20% acetic acid, 19% H₂O, 60% ethylene glycol. ×250.

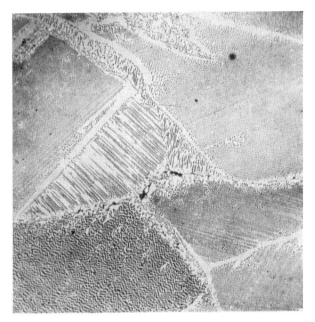

FIG. 3.9a

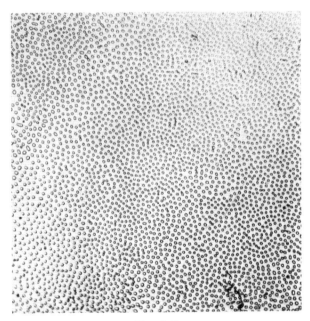

FIG. 3.9b

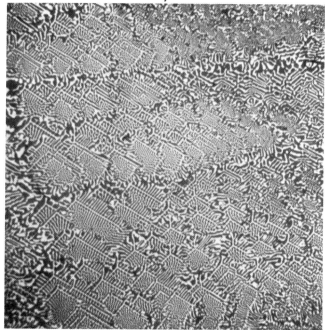

FIG. 3.10. This microstructure illustrates the dendritic or "Chinese script" form of eutectic structure. The alloy is Nb–21 w/o O and represents the eutectic composition between the phases NbO and NbO$_2$. The eutectic colonies can be seen at this magnification as long columnar-type growths. The regularity of structure is distorted at the boundaries between the eutectic colonies, and it is these lines of distortion that make it possible to see the colony structure.

Unetched. ×250.

FIG. 3.9. Comparison of the low magnification picture in (a) with the higher magnification view of one of the areas in (b) shows that this eutectic structure is made up of a bundle of long, needlelike, parallel rods. The alloy is representative of the eutectic composition (4% C) in the Cr-C system. The lower magnification view shows colonies of eutectic of Cr solid solution and the carbide, Cr$_{23}$C$_6$. (Figures courtesy of Dr. J. Westbrook, General Electric Research Laboratory, Schenectedy, New York.)

Unetched. (a) ×80. (b) ×400.

continuous and therefore of singular crystallographic orientation. Fixed crystallographic relationships exist between the plate-shaped discontinuous phase and the matrix phase. When a strong or primary direction of heat flow exists, the platelets will be generally aligned in that direction. This is another way of saying that the platelets lie orthogonally to a freezing front, if one exists.

Other two-phase configurations can develop by eutectic freezing. Sometimes the growth anisotropy of the discontinuous phase produces rod-shaped, parallel particles in a matrix (Fig. 3.9). The discontinuous phase can show the dendritic feature in its growth pattern shown in Fig. 3.10. The conditions that lead to these relatively unusual growth morphologies are not well understood.

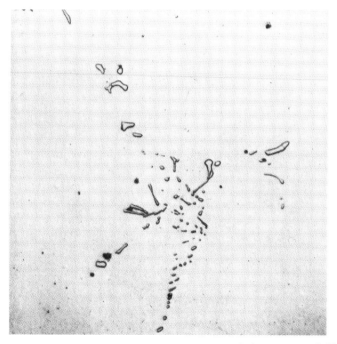

Fig. 3.11 The 0.15% Cb in this steel is the cause of the eutectic of CbC and austenite formed between the primary dendrites of austenite. The example shown illustrates a degenerated form of eutectic wherein the shape and orientation of the carbides have lost most of their distinctive nature. (Steel of 0.30% C, 2% W, 1% Mo, 0.5% V, 0.15% Cb.)

Unetched. ×500.

When the formation of eutectic is a secondary process—that is, predominantly primary crystallization followed by secondary crystallization of a small amount of remaining liquid—the lamellar symmetry degenerates as in Fig. 3.11. Under conditions of small volume of eutectic liquid and rapid freezing, one of the eutectic phases can form as a simple band around the primary growing grains as in Fig. 3.12. This seems to be a situation when growth of one of the eutectic phases on to the existing crystals substitutes for independent nucleation. The terms "divorced," "segregated," or "degenerate" are applied to these eutectic structures. A highly divorced eutectic structure cannot without additional information be distinguished from peritectic crystallization.

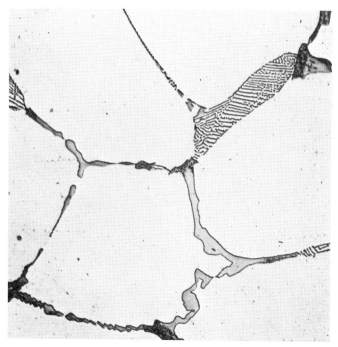

Fig. 3.12. The character of the eutectic of Mg solid solution and Mg₅Th depends very much on the rate of freezing. At slow rates the eutectic is lamellar, and at fast rates strongly oriented growth onto Mg dendrites produces a segregated eutectic which is essentially a band of the intermediate phase. The microstructure shown represents the structure of a hypoeutectic alloy containing 20% Th which has experienced an intermediate rate of freezing and as a result develops both lamellar and segregated (divorced) eutectic between the arms of Mg dendrites. Unetched. ×500.

In unusual or infrequent instances, a small volume percent of primary crystallization product can so dominate the freezing process that the primary crystallization particles feed on or grow from the eutectic melt leaving a majority matrix phase as a continuum to fill space. Nodular cast irons seem to be an example of this (see Fig. 3.13).

Another degeneracy in eutectic crystallization shows a discontinuous phase which possesses no dominant parallel features as shown in Fig. 3.14 (see also Fig. 3.3). Flake graphite in gray cast irons represent a complete degeneracy of the classic eutectic structure. Most gray cast irons are hypoeutectic, so that primary crystallization precedes the predominant (by volume) secondary crystallization of graphite and

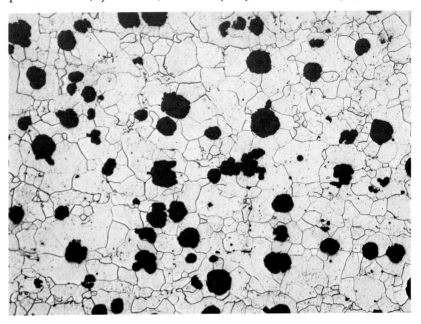

Fig. 3.13 A hypereutectic gray cast iron that has been inoculated with a powerful sulfur scavenger so that the primary crystallization product is spheroidal graphite. The eutectic melt freezes by formation of a graphite–austenite duplex structure. But in this case, the eutectic graphite simply grew on to the existing primary nodules of graphite leaving the austenite to crystallize independently and fill the remaining space as a matrix. Below the solidus, the austenite transformed to pearlite, but during the cooling, the carbide decomposed or graphitized depositing more graphite on the nodules and leaving a purely ferritic matrix. Thus, the nodules seen represent three growth processes that are superimposed. Etchant: nitol. ×100.

austenite. Except in rare cases (Fig. 3.6), the newly forming austenite grows onto the primary dendrites of austenite. The graphite nucleates independently into the form of what appears to be curved flakes that have no semblance of parallelism (see Fig. 3.15).

FIG. 3.14. Cast structure of a B-C alloy containing 42.5% C. The micrograph shows large acicular primary dendrites of graphite in a matrix of a eutectic of graphite and boron carbide (B_4C). The eutectic itself is an example of a non-lamellar structure, the graphite platelets being divergent or convergent to their neighbors.

Unetched. $\times 100$.

Peritectic Crystallization

In certain phase relationships, the primary crystallization phase can react with the surrounding melt to form a new solid phase. In a binary system, this is one of several possible invariant reactions—that is, the compositions of the reacting and product planes and the temperature are specified for equilibrium conditions. In general, equilibrium conditions do not obtain in normal freezing times. The reaction need not

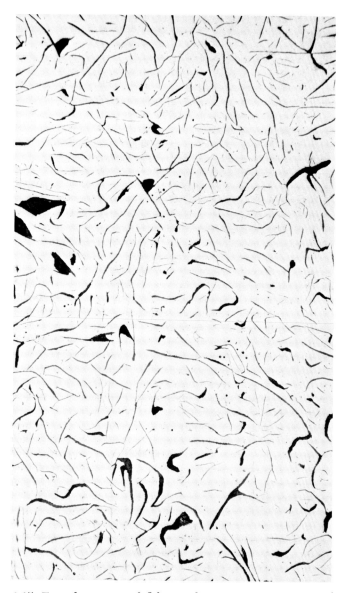

FIG. 3.15. Typical structure of flake graphite in a gray cast iron. The flakes appear to be separate entities, unsymmetrical both in shape and arrangement. Actually, the graphite growth from eutectic crystallization is petal-shaped, which in plate section appears as flakes. Moreover, the petals are in a roselike configuration in space. Some of the blocky-shaped traces of graphite in the microstructure reflect the chance oblique plane of section to one of the petals. This specimen is slightly on the hypoeutectic side of the eutectic composition.
    Unetched. ×100.

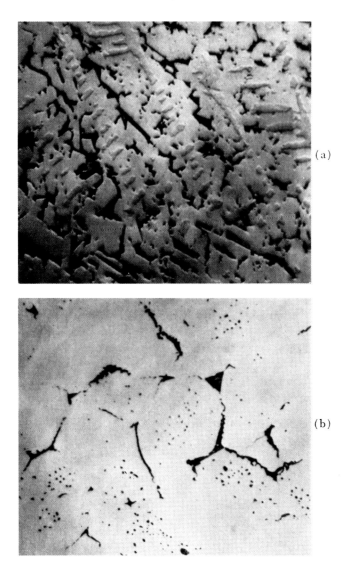

(a)

(b)

FIG. 3.16. The Zr-Mo system has one intermediate phase formed by peritectic crystallization involving primary dendrites of Mo solid solution and melt. In (a) is shown a classic case of a nonequilibrium cast structure with remnants of primary dendrites of Mo surrounded by the peritectically-formed ZrMo₂ phase and interdendritic eutectic of Zr solid solution and ZrMo₂. By annealing for many hours below the solidus temperature for this alloy (59% Mo), diffusion processes eliminate the metastable Mo, giving the microstructure shown in (b) which is essentially large grains of ZrMo₂ and some intergranular vestiges of the eutectic.

Etchant: (a) and (b). 20% HF, 20% HNO₃, 60% glycerin. ×600.

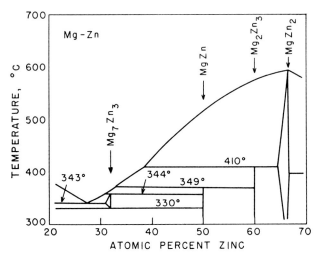

FIG. 3.17. Partial phase diagram of the Mg-Zn system.

FIG. 3.18. Micrograph (a) is a very unusual structure wherein (because of a series of incompleted peritectic crystallization processes) four phases are coexistent in a binary alloy. This is an alloy of magnesium containing 73% Zn (equiatomic composition). The phase diagram shows three peritectic processes:

$$Liquid + MgZn_2 \rightarrow Mg_2Zn_3 \qquad at\ 410°C$$
$$Liquid + Mg_2Zn_3 \rightarrow MgZn \qquad at\ 349°C$$
$$Liquid + MgZn \rightarrow Mg_7Zn_3 \qquad at\ 344°C$$

In its original solidification history, primary dendrites of $MgZn_2$ reacted with the melt so sluggishly that only the Mg_7Zn_3 phase could be recognized with certainty. By reheating to above the liquidus and slow cooling to 380°C and holding there for 15 minutes, the first peritectic process was initiated. The specimen was then quenched to room temperature and reheated to 335°C and held for 3 days. This permitted the second two peritectic processes to begin. The total time was insufficient to complete all processes that ultimately result in the single phase structure of the intermediate phase MgZn shown in micrograph (b). This end point requires prolonged annealing.

Etchant: (a) 4 gm picric acid, 1% H_3PO_4, 99% ethyl alcohol. ×150. (b) 1 gm I, 100 ml ethyl alcohol. ×150.

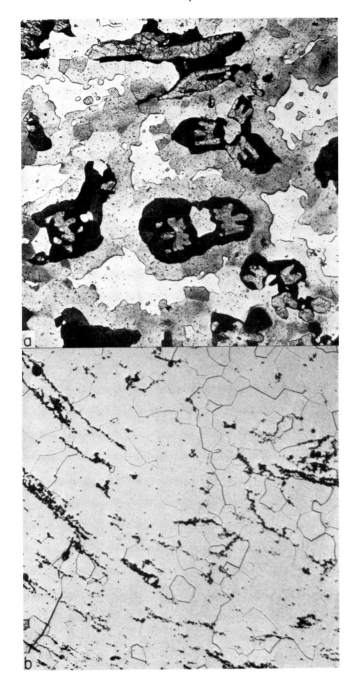

involve only stoichiometric proportions of reacting phases. Either excess primary crystallization phase or melt can coexist with the product phase below the peritectic temperature.

Since the reaction occurs at the solid–liquid interface and the rate of the reaction is limited by diffusion between the two solid phases, the peritectic reaction is often incomplete and stifled by cooling to lower temperatures where interdiffusion is essentially stopped.

The combination of events involving an incomplete peritectic and a divorced, subsequent, nonequilibrium eutectic cannot be directly inferred from Fig. 3.16(a). However, subsequent annealing restores the diffusion process to action that permits approach to equilibrium conditions. The change in proportion of phases showing in Fig. 3.16(b) indicates the existence of a peritectic process.

On occasion, the existing solid–liquid interface is not appropriate to the rapid nucleation of a new phase. In this case, the liquid may cool without change until it reaches some lower critical temperature where another form of crystallization can occur more easily. In this instance, a phase will be missing in the as-cast structure. A case in point occurs in the magnesium–zinc system. A portion of the phase diagram is shown in Fig. 3.17. In the normal course of solidification of a melt of initial composition at 50 a/o Zn, the MgZn phase is usually absent. This led to a controversy for a number of years over its actual existence. The absent phase can be made to appear by reheating and holding at some temperature in the (melt + MgZn) phase field. By stepwise annealing for carefully regulated periods of time, it is possible in this alloy system to induce the formation of four phases and prevent the disappearance of any of them. Such a four-phase structure is illustrated in Fig. 3.18(a). With further annealing below the solidus temperature, this structure would revert to the single phase dictated by the phase diagram [Fig. 3.18(b)].

Liquid Immiscibility

Complete melting—the fusion of all solid phases—does not of itself assure a homogeneous condition even with vigorous stirring. Heterogeneity among liquids can exist and gives evidence of this in the subsequent frozen state. The various micro- and macrostructural features can be discussed with reference to the phase relationships illustrated in Fig. 3.19

In the system illustrated by Fig. 3.19(a), homogeneity can be achieved at a sufficiently high temperature, but in the equiproportion

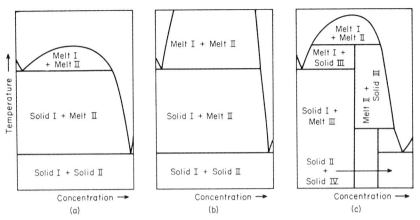

Fig. 3.19.. Diagrammatic illustrations of various phase equilibria involving immiscible liquids.

range, immiscibility develops on cooling and before any crystallization begins. The degree of physical separation depends on the rate of freezing. When the freezing rate is slow, there can be a continuing gravity separation of the two liquids. When the freezing rate is fast, the minor phase is trapped in the growing network of solids. Its shape, however, will have already been established in the liquid-liquid state. The liquid has no crystal character and directional heat flow cannot dominate surface tension; therefore, minor phase regions of liquid result in globules and this shape is preserved in the cast structure as seen in Fig. 3.20. It is even possible to suppress the precipitation of a second liquid by rapid quenching, for even a liquid must be nucleated. In such a case some crystallization operation dictated by lower temperatures will supplant the normal process.

In Fig. 3.19(a), melt I has a limiting composition and temperature at which it discontinuously precipitates solid until its composition has been modified to that of melt II. Since the temperature and compositions of phases participating are uniquely defined, this is one of the invariant processes in binary alloy systems and goes under the technical name of monotectic. The liquid separation and crystallization character of a monotectic crystallization process are illustrated in the micrograph of Fig. 3.21.

Figure 3.19(b) illustrates an alloy system which differs from Fig. 3.19(a) only in degree. Here, there is no possibility over a broad range of average compositions of obtaining a homogeneous melt. A

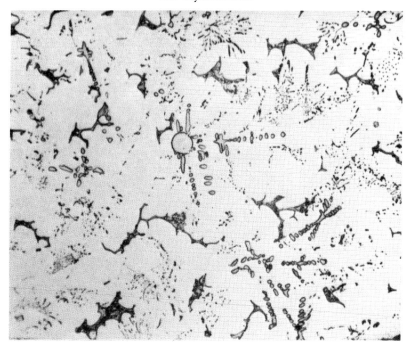

Fig. 3.20. This is the rather complex cast structure of an alloy containing 50% Fe, 35% Ti, and 15% O. The structure contains predominantly the intermetallic compounds TiFe$_2$ and TiO. A small amount of the ternary oxide phase Ti$_3$Fe$_3$O is indicated by X-ray diffraction but it cannot be specifically identified in the microstructure. The globule of TiO in the center of the micrograph indicates that a liquid immiscibility exists in this ternary system. The turbulence of the arc-melting operation prevented complete gravity separation of the two liquids, although some occurred as judged by the characteristic gold color of the TiO suboxide which showed on the top surface of the as-cast button in contrast to the cross section's shiny silvery appearance. During freezing, further primary crystallization of TiO occurs in the form of dendrites. The major phase, TiFe$_2$, crystallizes at a secondary stage. These are obviously two further stages of crystallization, one producing a coarse eutectic of TiFe$_2$ and TiO and a final one producing a very finely dispersed structure which is probably a ternary eutectic involving three solid phases. The Ti$_3$Fe$_3$O phase identified by X-ray diffraction is probably incorporated in this final crystallization product.
Etchant: 2% HF, 3% HNO$_3$, 95% H$_2$O. ×750.

gravity separation of liquid will occur at all temperatures up to the boiling point of one or another. Metastable dispersions can be obtained by mechanical turbulence but can only be retained by the most vigorous freezing rates. Where the engineering behavior of such phase mixtures is desired, the preferable process of preparation is by the

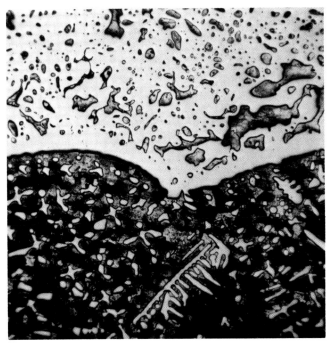

Fig. 3.21. This is the microstructure of a Cu–50% Pb alloy frozen in the crucible. Above 1000°C, this composition is one homogeneous liquid. Below this temperature, the alloy separates into two liquids and gravity forces the formation of a two-layer system with the heavier Pb-rich liquid on the bottom. At the monotectic temperature (953°C), the Cu-rich liquid contains 41% Pb and the Pb-rich liquid, 7.4% Cu. The monotectic crystallization involves the formation of massive solid Cu and the enrichment of the remaining liquid caught between the Cu dendrites to 92.6% Pb. At this point the molten lead layer is covered with a crust of solid copper populated with interdendritic pockets of Pb melt. With further cooling, the Pb melt layer rejects dendrites of Cu until it freezes as almost pure Pb at 327°C. The Pb melt trapped in the massive solid Cu does the same, but the Cu crystallizes on the walls surrounding the pocket.

Etchant: 15% HNO₃, 15% acetic acid, 60% glycerin. ×50.

infiltration of the lower melting liquid into a porous sintered body of the higher melting component at some temperature below the monotectic temperature.

The behavior of a third component in a liquid immiscible system merits some discussion. The lead-zinc system is of the form of Fig. 3.19(a). At 800°C, the two metals form homogeneous liquid solutions in all proportions. When silver is introduced as a third component, the miscibility condition is unchanged and a three-component homogeneous liquid is possible even with a very large proportion of silver. However, on cooling into the lead-zinc immiscible liquid field, the silver partitions almost completely to the zinc melt which separates out and floats to the top of the lead. This is the nature of the Parkes process for separation of silver from smelted lead. An appropriate amount of zinc is dissolved in the silver-bearing lead, and, when on cooling, the zinc liquid separates carries with it the silver values which are later recovered by distillation of the zinc.

There is another type of crystallization process involving a physical mixture of liquids by which at a critical temperature both liquids are constrained to precipitate the same solid species. This is one of the permissible invariant processes in binary systems and goes under the name of syntectic. The phase relationships are illustrated in Fig. 3.19(c). The subsequent solidification histories of the residues of melt I and melt II can lead to complicated nonequilibrium structures. The identification of the existence of a syntectic process can best be established by holding above the invariancy temperature until a gravity separation of the two liquids occurs and then demonstrating that the primary crystallization phase in each is identical.

Metastable Crystallization

In discussing the peritectic crystallization process it was pointed out that the crystallization of a phase can be suppressed under certain circumstances. This metastable condition can be reversed and the generation of the phase activated by appropriate reheating. Conversely, an intermediate phase can crystallize at an average chemical composition where its occurrence is precluded by the phase diagram. This anomolous crystallization can also be reversed by appropriate post heat treatment, the result in this case being resolution of the metastable phase. An example of this is illustrated in Fig. 3.22(a) and (b); in (a) is shown the envelopment of the austenitic grains by a segregated eutectic of alloy carbide. In the absence of the appropriate polycomponent phase diagram, the instability of the eutectic could only be tested by re-heat treatment to temperatures approaching the solidus. In this case, the carbide networks dissolved before re-melting occurred. Had the eutectic been truly stable, the heat treatment would have simply tended to spheroidize the carbide shapes.

Nonequilibrium compound crystallization originates from the same process that leads to coring in more widely miscible solid solutions. In fact, the two structure types often coexist. As described in the section on coring, the great disparity between homogenization rates

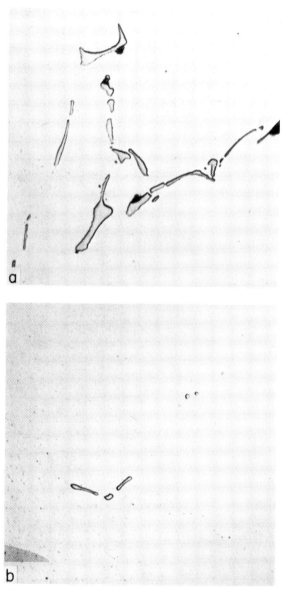

FIG. 3.22. The two micrographs are taken from an alloy steel containing 0.3% C, 2% W, 1% Mo, 0.5% V, and 0.15% Cb. Micrograph (a) represents the as-cast state and shows an interdendritic distribution of a massive columbium carbide (Cb-rich CbC). This carbide appears in its present location as a result of eutectic crystallization of alloy-enriched residual melt. The nonequilibrium nature of this carbide phase is demonstrated by the results of high temperature annealing at 2500°F shown in (b). In 2 hours at temperature, the carbides have been almost completely taken into the solid solution which at this alloy level and at this temperature should be homogeneous, single phase austenite.

(a) and (b) Unetched. ×1000.

in the solid and liquid states leads to the condition where the remaining liquid at some stage of crystallization is richer in alloy and the average composition of the solid phase is leaner than equilibrium dictates. If the cooling process could be halted at this point and solid state diffusion, which is the rate controlling process, allowed to correct the incompatible alloy distribution, then the normal phase relationships would be obtained in the cast structure. However, practical considerations rarely permit this and the over-rich alloyed liquid continues to cool. The dictates of phase equilibria are not relaxed because of these deviations; in fact, they are responsible for what follows. The phase diagram for the system may indicate that liquid of this alloy composition, irrespective of its metastability in the present context, will crystallize at some lower temperature by a eutectic or peritectic process. The cast structure will therefore exhibit the appropriate end product.

The occurrence of metastable eutectic, to choose a particular species, is a frequent dilemma in the delineation of new phase diagrams. This effect gives a fallacious indication of the maximum terminal solid solubility—indicating a magnitude much less than actual. It is, therefore, quite incorrect to set the maximum solid solubility as the composition in as-cast structures at which eutectic cannot be observed. For surety of judgment, it is necessary to anneal cast specimens for prolonged periods of time above the established eutectic temperature. This method is preferable when practical because if the eutectic is metastable, the re-melted liquid will gradually disappear and can be observed to do so. The process of resolution is accelerated by prior deformation. This serves to distort existing composition gradients. By so increasing the steepness of the composition gradients, the diffusion processes of resolution are temporarily accelerated. Sometimes it is found that solution of metastable compounds is practically feasible only in the more dilute alloys. Using these alloys to define a portion of the solid solubility boundary, the results can be plotted in a fashion: logarithm of the solid solubility versus the inverse of the absolute temperature. These curves are usually linear or nearly so and can be extrapolated to the established eutectic temperature to provide a reasonable approximation of the maximum solid solubility. The plot of the solidus temperatures of these same homogeneous alloys in the same fashion can provide a second extrapolated estimate for cross checking since the maximum solid solubility in a phase diagram is the threefold intersection point of the solidus boundary, the solid solubility boundary, and the horizontal eutectic line.

Nonequilibrium solidification problems are usually magnified with increasing size of ingot. When the primary dendrites assume macro-dimensions or when inverse segregation or columnar solidification fronts displace alloy-rich liquid large distances, homogenization is literally impossible. At best, the metastable compound remains as distributed inclusions or, at worst, they induce hot shortness and unforgability.

Coring in Cast Structures

An examination of almost any phase diagram will show that the composition of the first solid crystallizing from an alloy melt differs from that of the remaining melt. Moreover, with continuing crystallization the composition of the solid depositing on solid solution nuclei changes progressively. Thus, if the phase diagram specifies a falling liquidus curve or surface, chemical analysis of the fully developed dendrite from center to outside will present a curve of continuously increasing alloy content. When the liquidus curve or surface rises, the converse is true.

This nonequilibrium pattern of alloy microdistribution is the result of the indigenous disparity between the liquidus and solidus and of the inequality of diffusion rates in the crystallizing solid and in the remaining melt. This inequality always exists and so, in principle, composition gradients always exist at the moment of complete solidification. In practice, it is often possible to eliminate apparent micro-inhomogeneity by subsequent annealing below and as near as feasible to the solidus temperature of the homogeneous alloy. By re-establishing maximum concentration gradients, combined thermal-mechanical treatments can be even more effective. Where diffusion rates are slow as in some very high melting point alloys or in some very low melting point alloys, simple homogenization anneals may be ineffective.

The rate of chemical attack by etching solutions is usually very dependent on solid solution alloy content with the result that the structure of nonhomogeneous dendrites has the graded light reflectivity shown in Fig. 3.23(a). The relatively darkened center (or relatively darkened periphery depending on the direction of the composition gradient and its influence on chemical attack) will follow the morphology of the dendrite as did the freshly crystallized alloy. The apparent distinction between center and outside is the origin of the term "coring" used commonly to describe the metallographic appearance of dendritic microsegregation.

Commonly, the rate of chemical attack increases sharply at some

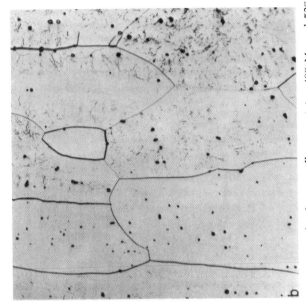

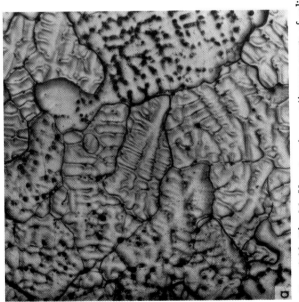

Fig. 3.23. The Mo–V system is a continuous range of solid solutions. The micrographs show an alloy containing 48% Mo and 52% V; (a) represents the as-cast structure and (b) is the same specimen after a 45-minute heat treatment at 1500°C. The etching differences brought about by the composition variation between the initial and final zones of crystallization have brought out the dendritic nature of the growth process by which the individual grains were formed. After diffusion has largely eliminated these zonal composition differences, the "cored" structure disappears.

Etchant: (a) and (b). 20% HF, 20% HNO₃, 60% glycerin. ×135.

critical concentration of alloy in solid solution. The result is an equally sharp change in etching intensity which makes it difficult to distinguish from a true grain or phase interface. One may easily be led to the conclusion that a two-phase structure has developed from the solidification process. Whether or not this is an illusion may be settled by

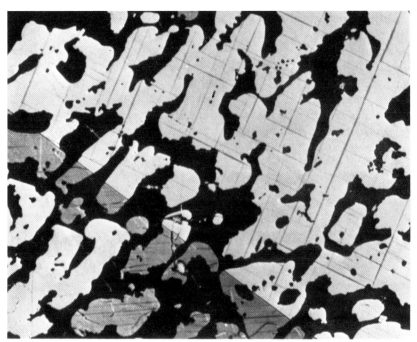

FIG. 3.24. In spite of its appearance, this is the as-cast single phase structure of a ternary intermediate phase occurring at the equiatomic position in the Ni-Ti-Si system. There is a sharp discontinuity in oxidation rate which makes for the appearance of two phases in the structure. But the single phase nature (apart from X-ray diffraction analysis) is betrayed by the long, growth-twin boundary traversing all discontinuities of etching. It happens infrequently that two adjacent crystallization nuclei bear a twin relationship to each other. Under these circumstances growth impingement produces a perfectly planar interface. In a more ductile material the same effect could be produced by recrystallization twinning as a result of thermally induced strains during cooling. But silicides are not very ductile and it is unlikely that sufficient deformation could be produced to initiate recrystallization. However, thermal stresses or metallographic preparation has caused a certain amount of mechanical twinning which interestingly enough seems to be localized to only one of the regions of the cored structure. (Figure courtesy of Dr. J. Westbrook, General Electric Research Laboratory, Schenectedy, New York.)

Heat tinted. ×100 PL.

examination of the structure after subsequent thermal or thermal-mechanical treatments. When the coring effect disappears on subsequent annealing as shown in Fig. 3.23(b), the decision is easy. However, when diffusion at practical temperatures and times is slow and when brittleness precludes mechanical working, other signs must be sought (see Fig. 3.24, for example).

If the hypothetical second phase appears to occupy more than about 10% of the area of the structure, one may reasonably expect to see a superposition of the diffraction lines of two phases in a powder, X-ray diffraction pattern. Figure 3.25 illustrates a different case where grain growth has created a network of grain boundaries superimposed and clearly unrelated to the pattern of coring. In spite of the persistence of the pseudo, two-phase structure, the existence of only one phase is a necessary conclusion. Cold work and recrystallization where possible can generate the same effect and permit the same conclusion.

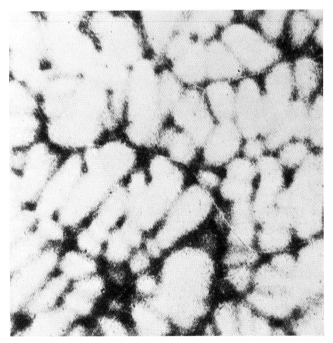

Fig. 3.25. As-cast structure of a V–10% Cu alloy. That this is actually a "cored" condition and not a poorly resolved two-phase structure is demonstrated by the grain boundaries which during cooling after freezing have migrated so that they now traverse both alloy-rich and alloy-poor zones.

Etchant: 20% HF, 20% HNO_3, 60% glycerin. ×250.

In both cases it is apparent that the concentration gradients in a solid solution structure are not effective barriers to the migration of grain boundary interfaces.

Frequently, coring can be immediately distinguished from a true, two-phase structure by the gradual transition in etching intensity from the light-to-dark etching regions. The boundary between a solid solution dendrite and an enveloping second phase as developed by a segregated eutectic or peritectic reaction processes is invariably sharp, consistent with the discontinuity in chemical composition which exists at the interface.

Incipient Melting

A previous section discussed the recognition of crystallization of a solid from a prior liquid state. The converse of this is the nucleation of liquid in a polycrystalline solid on heating. The subject is one of both practical and academic interest. The detection of the first indications of melting in a previously heated specimen represents a well-used technique for delineating the solidus temperatures of a phase diagram. As a method it is more convenient than precise. A specimen is heated to successively higher temperatures and examined metallographically for signs of melting after each heating cycle. The temperature of the first indication of melting and the previous temperature, where none was observed, provides a bracket of the solidus temperature. The accuracy of interpolation is no better than the temperature span of the bracket. Repetition of this procedure with alloys of successively increasing alloy content permits the construction of a smooth curve through the data, which increases the accuracy of each interpolation.

From a practical viewpoint the indications of incipient melting can be the basis of post-mortem rationalization of certain mechanical failures. Of course, by the same token, they can be the rewards of wise advance exploration by which premature and unexpected failures are prevented. The hazards of incipient melting are twofold—at the time of the event and later when again at room temperature.

If a small amount of liquid is nucleated at a hot working temperature, there is a strong likelihood of cracking. This condition of hot shortness can exist even though the liquid phase represents only a few per cent by volume of the total structure. But a small volume of liquid strategically disposed in envelope form about the grains of a structure can weaken it to a degree quite disproportionate to its amount. Of course, a structure whose grains are completely enveloped by liquid cannot sustain any significant stress and typical intercrystal-line failure results. Hot shortness can occur even under the circumstances that the liquid does not envelope each grain but exists as isolated intergranular pockets. This has been explained[*] in terms of the role of interfacial energy in fracture.

Very simply, the stress to propagate an existing crack is proportional to the change in surface or interfacial energy associated with the propagation of the crack into sound material. An island of liquid at the juncture of three grains constitutes a crack. The crack may be regarded as one filled with nonload supporting but surface wetting liquid. The propagation of the crack along a grain boundary reflects the extension of the two solid-liquid interfaces at the sacrifice of the grain boundary surface. This represents per unit area of crack formation an expenditure of energy designated as:

$$2\gamma_{S-L} - \gamma_B$$

where γ_{S-L} = interfacial energy between the solid and the liquid and γ_B = grain boundary surface energy. There is substantial evidence that γ_{S-L} is frequently not much more than one half of γ_B so that the expenditure of energy in propagating a crack in the presence of a liquid can be very small. It can also happen that $\gamma_B > 2\gamma_{S-L}$ in which case the melt completely surrounds each grain and almost no force is required to cause rupture. In fact the piece literally crumbles into individual grains as in Fig. 1.2.

Trace elements are an important factor in the industrial occurrence of hot shortness. Books summarizing phase diagrams contain many instances where a fraction of 1% of an alloying element reduces the solidus temperature by several hundred degrees. See for examples the phase diagrams of the systems Fe-B, Cu-Bi, Ni-S, Cr-Ce, Mo-C.

The nucleation of incipient fusion must be very rapid. The most suitable extrusion temperatures for high strength aluminum alloys are below but close to the solidus temperatures. Unless the ram speed of the press is quite slow, the extrusion will contain hot shortness cracks. These derive from temperature surges in the region within and immediate to the extrusion die because of the heat of deformation. When deformation is very rapid, the heat is generated in what is essentially an adiobatic system and significant temperature rise can result. Thus, high ram speeds can lead to internal temperature surges, generation of incipient fusion, and hot cracking from frictional tensile stresses in the very short time that any unit volume of metal is in the die zone.

Even if the thermal cycle which produces incipient melting is not

[*] R. Eborall and F. Gregory, *J. Inst. Metals* **84**, 88–90 (1955–1956).

concurrent with hot working, the resultant structure at room temperature is permanently deficient in mechanical properties, particularly ductility and toughness. The magnitude of property deterioration depends on the nature of the cast structure resulting from the ultimate solidification of the liquid. The eutectic envelope structure shown in Fig. 3.26 is a case in point. This micrograph is representative of a tool steel which in the as-cast state is quite brittle. Only by careful forging and heat treatment can the insoluble carbides be spheroidized and adequate toughness gained. The selective intergranular disposition of the eutectic structure resulting from the incipient melting restored almost the full brittleness of the original as-cast state.

FIG. 3.26. This specimen of M2 high speed steel has been heated to >2300°F and quenched. The microstructure demonstrates that the heat treatment temperature was above the solidus and as a result a small amount of liquid formed which enveloped the austenitic grain boundaries. The melt itself freezes by a eutectic crystallization involving complex carbides and austenite. The structure of the eutectic is barely resolved in the microstructure. The rounded austenite grains transform largely to martensite. Although a substantial amount of austenite is retained, this cannot be appreciated from the microstructure because the primary indication of this is the existence of austenite—austenite grain boundaries which in the present case are eliminated in the envelopment by the melt. (Steel of nominal composition: 0.8% C, 4.0% Cr, 2.0% V, 6.0% W, 5.0% Mo.)
Etchant: 5 gm $CuCl_3$, 10% HCl, 90% ethyl alcohol. ×400.

The detection of incipient melting depends on the recognition of a cast structure which derives only from the previous thermal cycle. A comparison of Fig. 3.26 with Fig. 1.20 makes this point clear. In the initial condition, the grain boundaries of austenite were free of all systematic distributions of second phase. This was achieved by the previous thermal-mechanical history. In the original as-cast state, this steel possessed a structure which can be described as primary dendrites of austenite and an interdendritic distribution of a carbide-austenite eutectic. If this structure were reheated to above the solidus temperature, the eutectic would remelt and on subsequent cooling re-crystallize to a eutectic structure which is indistinguishable from the original cast state. Thus, in general, it is difficult or impossible to identify the occurrence of incipient fusion in as-cast specimens. The method is applicable only when, by thermal or thermal-mechanical treatment, the original as-cast structure is replaced or modified.

Although incipient fusion most frequently leads to the formation of pockets of liquid at three-grain junctions (stretching into regions of two-grain junction), this is not invariably so. On occasion, melting can occur in the interior of grains but probably this event begins first at grain boundary intersections. Since, in the middle of grains, no triangles of surface tension forces can be set up, the pockets of liquid will assume the spheroidal forms shown in Fig. 3.27.

The geometric shapes of the liquid pockets at three-grain intersections offer one of the direct methods of measuring the relative surface energy of the solid-liquid interface. The corner of each triangular shaped, erstwhile liquid pocket represents the point of balance of the grain boundary surface tension with the two solid-liquid interfacial tensions. The condition of equilibrium is defined as:

$$\gamma_B = 2\gamma_{\text{L-S}} \cdot \cos \frac{\theta}{2}$$

where θ = angle between the traces of the two converging liquid-solid interfaces. This angle is called the dihedral angle.

Again it is necessary to recognize the three-dimensional nature of the shapes of phases. Although each island of liquid is in thermodynamic equilibrium with the enclosing three grains, the arbitrary sections will give a spectrum of measured dihedral angles. The correct dihedral angle must be determined by statistical measurement. For a large number of measurements (perhaps 50–150), the true dihedral angle will be given by the peak of the distribution curve. The size of the sample is governed by the need to define clearly an optimum angle.

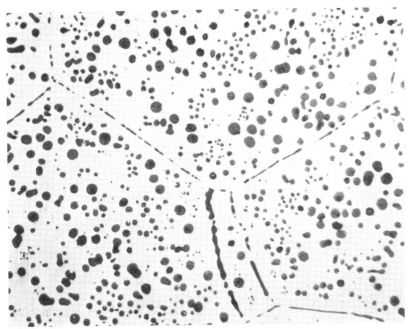

FIG. 3.27. On rare occasions melting can nucleate inside a grain as easily as at a grain boundary. This is particularly true when the interfacial energy between the melt and the solid is relatively high. This is indicated in the microstructure of a columbium alloy containing 10% Ti which has been heated to 2245°C. The areas of melt phase at the β phase grain boundaries are largely discontinuous which signifies a large interfacial energy. The pockets of melt forming inside the β grains having no crystalline character have no preferred directions of growth and no discontinuous interfaces to establish dihedral angles with. They therefore assume spherical geometry. A Zr–15% Pt alloy shows similar tendencies. [See E. G. Kendall, C. Hays, and R. E. Swift, *Trans. AIME* **221**, 445–452 (1961).]

Etchant: 1% HF, 3% HNO₃, 96% H₂O. ×120.

The tendency for a liquid to envelope completely the grains of a polycrystalline structure is governed both by the amount of liquid present and the ratio of γ_B to γ_{L-S}. If γ_{L-S} is less than $\frac{1}{2}\gamma_B$, i.e., $\theta < 60°$, then, even with a very small amount of liquid, the grains will be completely enveloped. If γ_{L-S} is very large so that the dihedral angle approaches 90°, even very large amounts of liquid will remain as isolated globules.

Columnar Cast Grain Structure

Most pure metals and many alloys when solidified under conditions of a continuous and steep gradient will produce a fibrous-appearing

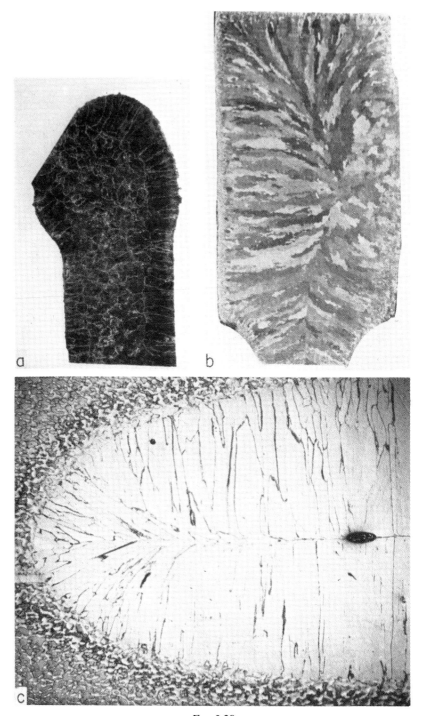

Fɪɢ. 3.28

cast grain structure whose directions of fibering are perpendicular to the casting surfaces. Under higher magnifications the individual fibers are revealed as parallel elongated grains whose major dimensions are many times either of the two other principal transverse dimensions; hence, the term "columnar grain" structure.

The columnar type of structure develops in the presence of steep thermal gradients. High rates of heat abstraction and well-defined directionality of heat flow usually go together. Hence, the structure is common in "chill" castings—in castings "chilled" in selective locations and in weldments. Not all alloys are prone to this type of structure even under conditions of rapid solidification. The columnar structure develops by the growth of crystallization nuclei from the mold surface inward toward the centerline of the melt. This directional growth will continue uninterrupted unless forestalled by crystals which nucleate ahead of the main solid-liquid interface because of local changes in the alloy composition of the liquid,* or by interference from similar solidification fronts advancing from other mold. faces. Both of these circumstances can be appreciated from Fig. 3.28.

The columnar grain assembly has a positive preferred orientation character. In the case of cubic metals, the axes of the elongated grains coincide with the [100] crystallographic directions. In this crystallographic direction, a crystal grows most rapidly. Accordingly, of the multitude of nuclei formed at the mold wall/casting interface, those accidentally oriented with the cube direction parallel to the direction of heat flow will grow very much faster than the others. In a very

* For more detailed discussion, see B. Chalmers, "Physical Metallurgy," pp. 272–277. Wiley, New York, 1959.

FIG. 3.28. Each of these macrographs reveals some aspect of columnar grain growth. In (a) representing a steel casting of AISI 1040, we see the sharp transition from columnar to equiaxed grain shape. The portion of the casting in (b) produced from an 18% Cr–8% Ni steel was adjacent to a riser (on the right but cut away). The axis of the columnar grains points to the principal direction of heat flow during freezing. The very thin rim of fine equiaxed grains characteristic of a strongly chilled surface may be seen. Macrograph (c) is taken from a spot weld in Muntz metal (60% Cu–40% Zn). The spot weld represents a very small region of fusion surrounded by the massive chill of the parent metal. The pattern of columnar growth forms a plane of juncture along which accumulates all of the shrinkage voids and freezing segregants. It is generally therefore a plane of serious mechanical weakness. Note that while the base material is two phase ($\alpha + \beta$ brass) the weld nugget is single phase β indicating the rapidity of the quench.

Etchant: (a) and (b) 5 gm FeCl₃, 10% HCl, 10% HNO₃, 5% H₂SO₄, 75% H₂O. ×0.7. (c) 10% NH₄OH, 90% H₂O. ×28.

short space, their minor lateral growths link up to form a solidification front sealing off the less fortunate nuclei from further growth. It is frequently possible to see in the cross section of a chilled casting, a thin surface layer of fine equiaxed grains, followed by a large number of short length columnar grains, in turn followed by many fewer which reach deep into the section of the casting. This sequence is representative of the spectrum of growth rates exhibited by the various surface nuclei.

The well-defined columnar structure of Fig. 3.28(b) is more common to the chill casting because the severe undercooling of the surface liquid leads to a very large number of crystal nuclei. When the rate of heat abstraction by the mold is less, so is the number of surface nuclei produced, the number favorably oriented for fast growth, and hence the number of columnar grains per unit area of solidification front. The term "columnar" can barely be applied in these circumstances because the transverse dimensions of the grains may not be so very much less than the longitudinal dimension.

Concomitant with the elongated grain shapes of the columnar structure are long, uninterrupted grain surfaces or (more properly) interfaces. These are favored locations for collection of impurities which concentrate in the last vestiges of liquid and for alloy-rich, late-solidifying liquid. The continuity of such imperfections jeopardize the mechanical properties of the casting.

The patterns of columnar grain structure identify the major directions of heat flow. Furthermore the relative depths of columnar grain penetration are a good guide to the relative rates of cooling or chill capacity at the various casting surfaces. The planes of juncture of columnar grain growth fronts have long been correlated with tendency to cracking during forging or rolling. Such a plane appears in Fig. 3.28(c) as a line replicating the original faying surface of the joint. Here the interlocking of nonparallel grains must lead to gross segregation and shrinkage or gas porosity.

Grain Size in Cast Structures

It is a mistake to think that everything important can be seen at high magnification. In fact, the view at high magnification can often be quite misleading. The grain size of the cast aluminum alloy shown in Fig. 3.29 is a case in point. At the two different magnifications the views are very different. One is led to believe that the grain size as seen at $\times 85$ is much smaller than the $\times 0.9$ magnification suggests.

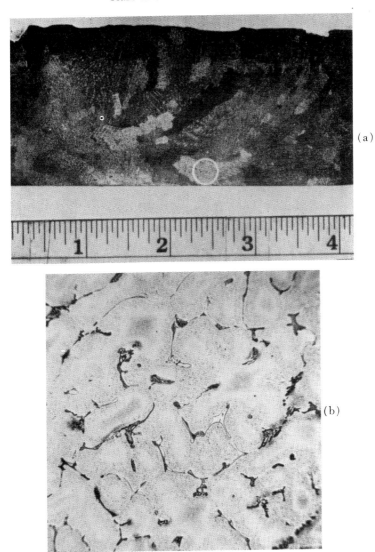

(a)

(b)

FIG. 3.29. Macrograph (a) and micrograph (b) represent two views of the same structure at different magnifications. The specimen was cut from a large ingot of a dilute Al-Cu alloy which was frozen very slowly. The grains as seen by eye (a) are very large, yet seen at ×100 magnification one would guess that the grain size is quite small. The view in micrograph (b) represents a planar section through the branch structure of a single dendrite, each arm of which is enveloped in eutectic. The apparent grains delineated by the eutectic rims are all of the same orientation. Note the cored structure in (b).

Etchant: (a) 5% HF, 95% H_2O. ×0.9. (b) 1% HF, 1.5% HCl, 2.5% HNO_3, 95% H_2O. ×85.

The low magnification view represents the true grain size in the sense that each grain represents the completeness and complexity of an individual dendritic growth. The dendrite in its final form represents several orders of branching and subbranching. Since, however, crystal growth is disciplined by preferred crystallographic directions, the major axes of the branches and subbranches have rigorous angular relationships with each other. Moreover, the [100] or cube direction is preferred in cubic metals so that branches are all parallel or orthogonal to each other and any arbitrary plane of section reveals the same planes in all branches.

Yet in Fig. 3.29(b), a fine grain pseudostructure is seen wherein each grain is enveloped by divorced eutectic. This view really illustrates the final distribution of eutectic liquid within the dendritic structure. In so doing the individual branches of the dendrite are physically separated and appear to be an assembly of different grains. This, of course, is not true. The grains are all of equivalent orientation but this can only be appreciated by the uniformity of etching seen at low magnification.

The intricacy of detail of the interdendritic eutectic distribution and its influence on light etching masks the true grain boundaries which cannot be seen at the high magnification. This is a valid instance of "not being able to see the forest for the trees." In this alloy the eutectic is nonequilibrium and with long annealing can be almost completely dissolved into the matrix solid solution. When this is accomplished, it is possible to bring out the true grain boundaries at higher magnifications.

The term "grain size" can on occasion have expanded meaning to include the size of other growth entities. A case in point is the eutectic cell size in gray cast iron shown in Fig. 3.30. The boundaries between

---

FIG. 3.30. The structure of gray cast iron involves a minority of primary dendrites of austenite (transformed into ferrite and pearlite) and a majority of eutectic of graphite and austenite (also transformed). Each eutectic colony is a nonlamellar, complex intertwining of the two constituent phases containing also a few primary dendrites of transformed austenite randomly oriented and superimposed on the eutectic structure. At the boundaries between eutectic colonies, such trace elements as sulfur and phosphorus collect and form part of the last melt to freeze. The etching contrast between the eutectic colonies and the impurity-rich boundaries delineates the eutectic cell size and shape. The eutectic cell size in gray cast iron is a factor in mechanical properties, just as grain size is a factor in other alloys. (Figure courtesy of Dr. H. Merchant.)
Etchant: 4 gm $CuSO_4$, 50% HCl, 50% H_2O. ×10.

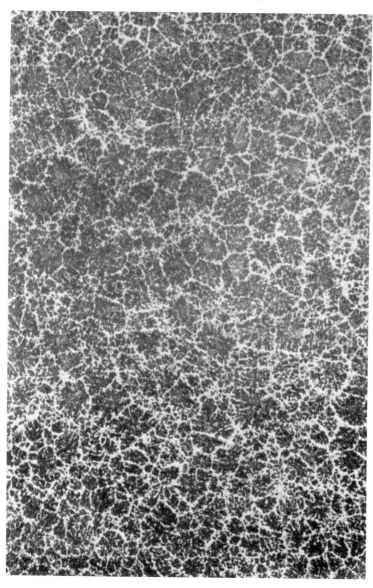

FIG. 3.30

eutectic cells etch selectively because of segregation of minor elements such as sulfur and phosphorus. Each eutectic cell contains an intricate, intertwining of austenite and graphite of which all components of austenite (and of graphite) are in rigorous orientation relationship.

Porosity in the Cast Structure

There are two important sources of porosity in a cast structure—shrinkage and gas. In most metals the volume change accompanying solidification is negative. Under ideal conditions the shrinkage space associated with solidification is progressively filled by the natural, one atmospheric pressure feeding from the reservoir of remaining liquid. Under conditions of dendritic solidification, the interlocking action of adjacent dendrites and merging dendrite arms may block off normal feeding. Liquid metal so trapped must leave a shrinkage void on solidifying. This void will in general assume the geometry of the space between adjacent dendrites. The long dimension of these voids is closely related to the grain or dendrite size. Finer grained castings yield commensurately finer porosity with less consequent degradation of mechanical properties.

In the molten state, many metals have an appreciable solubility for gases such as hydrogen, nitrogen, and carbon dioxide and very much less solubility in the solid state. The rejection of gas is therefore concurrent with the solidification process. When the melt is initially saturated with gas, the continuous evolution during solidification results in bubble shapes which are unmistakable.

When the melt is not initially saturated, gas will concentrate in the remaining liquid during solidification building up to saturation level. The porosity will appear as a concentration of voids in the center of the casting and particularly under a riser. The condition of saturation may well coincide with the state of interdendritic distribution of residual liquid. The resultant gas porosity confined between the arms of interlocking dendrites will form void shapes which are very much as shown in Fig. 3.31. In short, certain forms of shrinkage and gas porosity are indistinguishable simply because precipitating gas most easily emerges into existing shrinkage pores. The need for distinction is only to decide what types of corrective action in melt conditioning and casting design are necessary.

Gas porosity in a casting can also derive from the vigorous evolution of volatiles from surrounding mold materials. The pressure of

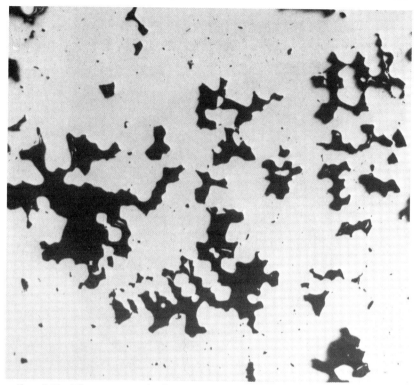

FIG. 3.31. This micrograph illustrates shrinkage porosity in the cast structure of a Cu-base alloy containing 5% Sn, 5% Zn, 5% Pb. These voids are clearly disposed between large dendrites and as such are termed interdendritic shrinkage porosity. The specimen was taken from a position directly beneath the riser of the casting, where this condition is most easily seen because of the slow rate of freezing and the large resultant dendritic growths. Shrinkage derives from the change in specific volume associated with the change of state from liquid to solid. The conversion of melt to solid about the periphery of each dendrite leads to void formation in these interstices. Ideally, these voids should be refilled with melt from some other location where a reservoir exists. Control of directions of freezing and the use of risers for feeding are major technical tools used to produce sound castings. Metallographic examination in the unetched state is an important basis for judgment of the effectiveness of casting design.
Unetched. ×50.

gas generated at the mold-casting interface can rupture the thin solid skin forcing a bubble into the mushy solidifying metal. Figure 3.32 illustrates a gas bubble trapped near the surface of a weld by such action.

In extreme cases, mold gases can literally bubble through a casting

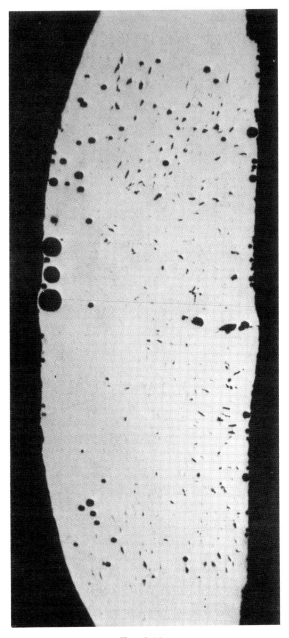

Fig. 3.32

during freezing, some of the bubbles being trapped beneath an upper flat surface of the casting. Again one must be very careful in attaching the blame, for poor chill and riser design with consequently inadequate feeding may produce the same effect.

It is well to remark at this point that both shrinkage and gas generated defects can range in scale from grossness that can easily be seen by the naked eye to sufficient fineness that requires ×50 magnification for proper observation. In general, fine, microporosity occurs in colonies—that is, large numbers in particular regions of a casting. The colonies themselves can be seen by the naked eye through the reduced reflectivity of the polished surface, but the shape and size of the individual voids must be examined under the microscope. The actual size of pores and voids is best judged from the as-polished, unetched state, provided that cutting abrasive such as diamond or sapphire is used for final finishing. The etching process attacks the edges of pores much more vigorously than the general surface causing a considerable increase in their apparent size. Examining a porous region in the etched condition serves only the purpose of locating the individual pores with respect to the components of the cast structure.

While this discussion has chosen examples from the foundry trade, many of the defects produced by dissolved, entrapped, and generated gases are also found in fusion weldments. This should not be surprising since a fusion weld is really a form of casting. Voids and porosity from gas are more common in welds than shrinkage defects because of the strongly directional pattern of freezing and the more careful choice of welding rod alloys from the viewpoint of their freezing characteristics.

Fig. 3.32 Molten Al and its alloys have a large solubility for hydrogen while in the solid state the solubility is extremely small. On freezing, therefore, hydrogen picked up during melting is almost competely expelled. The precipitated hydrogen will grow into bubbles and if unimpeded will rise to the top of the melt and disappear. Nonideal freezing patterns, however, lead to bubble entrapment as shown in this macrograph.

The unetched view represents the weld zone between two plates of an Al alloy containing about 4% Mg. The joint was made by arc welding. The arc is capable of dissociating water vapor into its component elements and so hydrogen is readily available for easy absorption when the humidity is high or the moisture content of the welding fluxes is high (if such are used). Most of the gas bubbles tend to rise to the top of the melt. A few are trapped near the bottom by crystallizing dendrite arms. The nonspherical voids randomly distributed throughout the weld zone may be either interdendritic gas porosity or shrinkage porosity, or both. In this type of specimen and at this magnification they are not really distinguishable.

Unetched. ×12.

Crystallization from Nonmetallic Media

While most experiences with crystallization involve derivation from a melt, the metallurgist must also recognize that there are alternative processes using aqueous solutions, fused salts, and gaseous mixtures. These produce structures which have some characteristics similar to castings and some quite peculiar to the process. Under certain conditions of vapor deposition and electrodeposition from aqueous solution and fused salts, crystallization is clearly dendritic (see Fig. 3.33). Metal powders produced by various electrolytic processes are generally dendritic, even though the particle sizes are small enough to require classification by fine screens.

As with solidification from a melt, the dendritic growth form can be retained in relative isolation. But in the growth of a dense single phase structure, mutual impingement imposes shape limitations and

Fig. 3.33. Scanning electron microscope view of the dendritic formation of tungstic oxide in air. A tungsten filament was resistance-heated to incandescence in air. The reaction with air produced an oxide formation. The growth of the oxide on the tungsten metal surface produced whiskerlike growths that at higher resolution reveal a classic dendritic form. (Figure courtesy of R. G. Wibel.)
×1000.

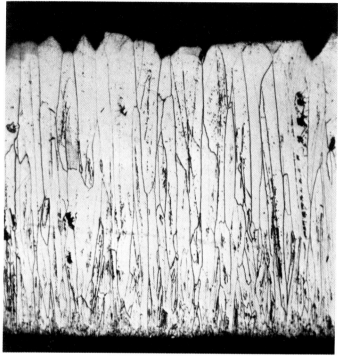

Fig. 3.34. Structure of electrolytically deposited iron illustrating the well-defined columnar growth which can develop by crystallization from an aqueous solution. It would be difficult on the basis of appearance only to distinguish this microstructure from a chill casting of the same metal.
   Etchant: 5 gm CuCl₃, 10% HCl, 90% ethyl alcohol. ×50.

the dendritic form becomes obscured. Dense electrodeposits frequently assume the columnar grain structure shown in Fig. 3.34 which represents a condition of mutually restrained lateral growth. This structure is fundamentally the same as encountered in chill castings of pure metals and single phase solid solutions. As in chill castings, the columnar grains have preferred orientations with their long axes corresponding closely to certain crystallographic directions. Otherwise the long columnar grains possess random rotational orientation about their principal direction. The grain structure of the substrate surface can serve as points of oriented nucleation. It is common to observe grains of the electrodeposit which appear to be continuous with individual grains of the substrate. It is reasonable to expect that in many more instances, there is some crystallographic conformity between grains of the substrate and of the adjacent deposit governed by rules of selection similar to those responsible for Wid-

FIG. 3.35. The structure of Ni, electrodeposited from a Watts bath, can be almost unresolvable even at the ×1000 magnification shown. This extremely fine structure provides unusually high tensile strengths (100,000–120,000 psi) compared to wrought and annealed Ni. On annealing the as-deposited structure at temperatures of the order of 650°C, pronounced grain growth occurs.
Etchant: 5 gm KCN, 1% H_2O_2, 99% H_2O. ×1000.

manstätten precipitate arrangements.

The appearance of the as-deposited structure depends very much on the conditions of electroplating. Grain size variations are the most obvious. In Fig. 3.35 the grain structure is not clearly resolvable. These structures are generally much harder and stronger than the coarse, columnar deposits. The hard electrodeposits usually show striations and unresolvable dispersions. Since the choice of composition of the electrolyte, both minor and major additions, can profoundly influence the hardness of the deposit, it is thought that the dispersions signify the existence of finely distributed impurities of undefined nature. This is another of many cases where metallography provides only clues instead of solid answers. Here substantial hardness increases correlate with choice of plating bath, plating conditions, and a structure obviously populated by a foreign substance in finely particulate form.

FIG. 3.36. Structure of chemically deposited Ni taken from a large lump deposit on the wall of a plating tank. Precipitation nuclei are widely separated and growth of each is well advanced before impingement interferes with the concentric circle pattern of enlargement. The growth interference is very much as occurs in normal crystallization except that the interfaces are immobile and cannot assume a minimum energy condition. The concentric striations represent periodic variations in P content of the amorphous Ni resulting from periodic depletions of the immediately adjacent plating solution.

Etchant: 5 gm CrO_3, 100 ml H_2O, used electrolytically. $\times 100$.

But the identity of the substance requires another tool. The thought is that these impurities provide a form of dispersion hardening. The capability is considerable, for electrodeposits of ordinarily soft metals with only trace detectible impurities can reach as-plated hardnesses of more than 500 DPH.

The fine dispersions, high residual stresses, and ultrafine grain size are all conditions of thermodynamic instability and post heat treatment produces recrystallization to equiaxed grain structures and hardnesses close to those encountered in equivalent cast and wrought metal. Certain impurities such as sulfur in nickel, however, cause permanent impairment of ductility.

The catalytic decomposition of certain complex nickel phosphate solutions provides a rare opportunity to study the growth of an amorphous metal from aqueous solution. The deposit is dense and noncrystalline, although crystallization occurs rapidly on heating to temperatures as low as 300°–400°C. The composition of the deposit is essentially nickel with about 7–9% phosphorus. The dominant characteristic of the structure of the deposit is the system of concentric or parallel striations vividly illustrated in Fig. 3.36. These striations have been interpreted as periodic variations in phosphorus content and all of the other structural details derive from impingement of adjacent striation systems. The concentricity or parallelism of striations represents a record of growth. In Fig. 3.37 it can be seen that the deposit began at isolated points growing out radially. The impinge-

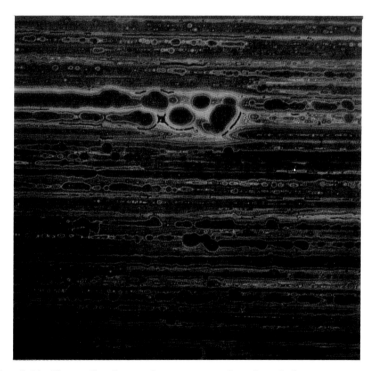

Fig. 3.37. Chemically deposited Ni on a steel surface looking down on the structure of the deposit in the plane of the surface. This structure reveals the point nucleation character of the deposit and the system of concentric and interlocking growth patterns. The ultimate development of the longitudinal growth lines reflects the parallel machining marks on the surface of the steel substrate. This is a very thin deposit and certain void zones still remain.
Etchant: 5 gm CrO₃, 100 ml H₂O, used electrolytically. ×100.

ment of concentric striation systems leads to blending into a plane growth front and the evolution of a single system of parallel striations. This can be better appreciated from the section view in Fig. 3.38. Figure 3.38 also illustrates growth faults in the planar growth over a large surface. The infinite "throwing power" of this type of nonelectrolytic deposit can be seen by the completeness with which the deposit fills the space underneath the machining burr on the steel substrate.

As stated earlier the amorphous deposit is highly unstable and on

FIG. 3.38. Chemically deposited, amorphous Ni deposited on a rough machined steel surface. The precipitation of Ni begins simultaneously at all points on the steel surface, and the rate of growth perpendicular to the surface is equal at all points. Periodic local variations in the composition of the solution change the amount of P deposited with the Ni, which accounts for the parallel striations. The displacement of the striations reflects the relative initial displacement of points on the surface of the steel. The very great "throwing power" of this plating process is revealed by the completeness of deposition at the root of the machining burr.

Etchant: 5 gm CrO_3, 100 ml H_2O, used electrolytically. $\times 500$.

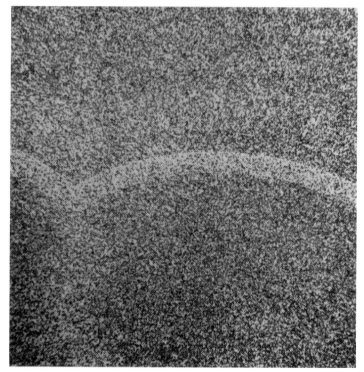

Fɪɢ. 3.39. Structure of chemically deposited Ni after heat treatment at about 800°C. As-deposited, this is a Ni-P alloy (~7% P) with a peculiarly striated amorphous structure. Being highly metastable, the anneal permitted the alloy to revert to its equilibrium state of a phase mixture of Ni (dilute solid solution) and Ni₃P which is seen here as a dispersion. The density of dispersion reveals that the striations in the as-deposited structure actually represent periodic variations in P content.

Etchant: 5 gm CrO₃, 100 ml H₂O, used electrolytically. ×500.

post heat treatment will crystallize to a phase mixture of nickel solid solution and Ni₃P. The density of distribution of phosphide particles in the heat-treated structure of Fig. 3.39 and the banded appearance of the zones lends support to the interpretation of the significance of the striations in the as-deposited structure.

Similar striated structures may be found in very fine grained electrodeposits which are unquestionably crystalline. It appears that periodic fluctuations in the concentration of co-deposited impurities is characteristic of plating processes although the impurity is clearly not always phosphorus.

CHAPTER IV

Solid State Transformations

Precipitation from Solid Solution

The morphologies of precipitants from solid solution are more diverse than from liquid solutions. These variations derive from factors which do not exist in noncrystalline supersaturated solutions. Chief among these are the grain boundaries and other interfaces in the supersaturated matrix and the almost universal ability to undercool and retain a state of metastable supersaturation for considerable time at temperatures many hundred degrees below the solvus boundary of the phase diagram. There are other important distinctions between crystallization from solid and liquid solutions. The volume rate of precipitation from solid solution is governed primarily by temperature, whereas from liquid solution the rate is governed by heat loss. Nucleation from solid solution may be either heterogeneous, preferring sites at interfaces, or homogeneous, having no apparent preference for site. Nucleation from liquid solution is almost exclusively heterogeneous— homogeneous nucleation being produced only by very elaborate laboratory precautions.

Precipitation processes are the basis for the control of mechanical and physical properties of alloys by heat treatment. Yet some of the most important stages of precipitation are invisible by ordinary metallographic techniques. However disappointing this may be, it is important to accept as a realistic limitation. In the condition of peak aged hardness in many alloys the small size of precipitant and their etching contrast will not permit resolution under the optical reflection microscope. For such studies recourse must be made to electron microscopy and X-ray diffraction.

Sites for Nucleation of Precipitation

The preference for certain sites as points of nucleation is inversely proportional to the degree of undercooling below the solvus boundary.

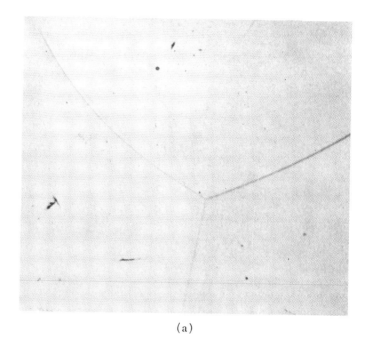

(a)

(b)

FIG. 4.1

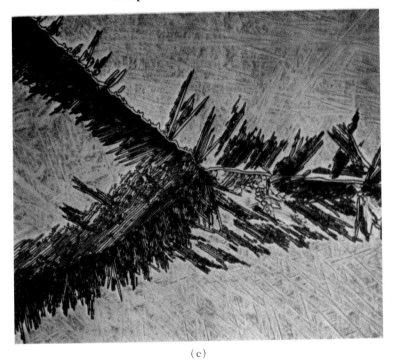

(c)

FIG. 4.1. The first two micrographs (a) and (b) represent very short successive intervals in the isothermal transformation of a Ti–11% Mo alloy from single phase β to an $(\alpha + \beta)$ phase mixture. The specimen had been solution-treated at 1000°C and then quenched to 700°C. In the first case the specimen was held for only 10 seconds before quenching to room temperature. The resultant structure shows retained β with thin, sharply defined grain boundaries. The second specimen quenched after a 30-second hold at 700°C, shows the first detectable onset of precipitation of α at the grain boundaries of the β phase. The β grain boundaries are etching wider and some spots of precipitate can almost be resolved. The black markings in the interior of the β grains are extraneous to the binary alloy and are probably hydrides. In a matter of a few minutes the grain boundary precipitate nuclei can grow to a size which clearly resolves their shape as shown in (c) for a Ti–3% Mo alloy isothermally transformed at 750°C for 5 minutes. In the interior of the β grains where precipitation has not relieved the supersaturated condition, the β has transformed on quenching to the martensite phase called α' because it is structurally identical to the equilibrium α phase.

Etchant: (a), (b), and (c). 20% HF, 20% HNO₃, 60% glycerin. (a) and (b): ×675; (c): ×250.

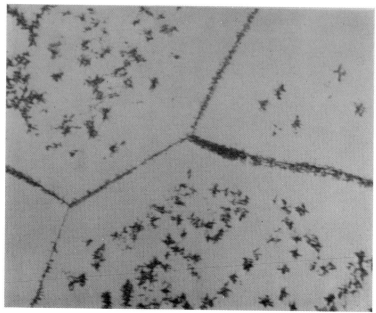

Fɪɢ. 4.2. This Ti alloy contains 7% Mo. It has been solution treated at 1000°C, quenched to 650°C where it was held for 15 minutes, and then quenched to room temperature. The structure is one illustrating precipitation of the α phase in the grains of the high temperature β-phase which is supersaturated with 7% Mo at 650°C. In point of time the platelet growths of α appeared first at the grain boundaries and later nucleated randomly throughout the interior of the β-grains. Etchant: 20% HF, 20% HNO_3, 60% glycerin. ×750.

At temperatures near the solvus, precipitation invariably begins at grain boundaries. Even before the precipitate shape can be resolved, the beginning of precipitation can be recognized by the increased etching rate at the grain boundaries. This preliminary sign is illustrated in Fig. 4.1. It is important to realize that the preference for nucleation at the grain boundary is not absolute and is only temporal. At the same temperature and later in time, precipitation will begin at a multiplicity of points in the interior of each grain. This is illustrated in Fig. 4.2. The sequence of events underlines the probability nature of nucleation. Statistically, nucleation at a grain boundary is favored by thermodynamic considerations. Yet a finite probability exists for nucleation elsewhere and so these other events will occur also, but later in time.

Other low energy interfaces such as twin boundaries, low angle or subgrain boundaries, and zones of lattice distortion such as deforma-

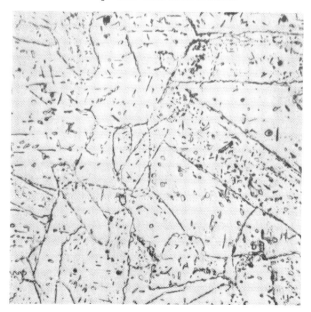

Fig. 4.3. Although the solid solubility of Cr in Co at 1000°C is about 35% and of Mo in Co about 20%, the ternary alloy of Co–25% Cr–10% Mo illustrated by this micrograph is in a two-phase field at equilibrium. The alloy has been annealed at 980°C for 64 hours (because of the slow rate of precipitation) and then water quenched. At this temperature, the low temperature hexagonal close packed polymorph of Co is stable. As is common with hexagonal close packed solid solutions, annealing twins are frequent. Twin boundaries being true interfaces are favored regions for nucleation of precipitation as illustrated in this micrograph by the discontinuous bands of the "sigma" intermediate phase along the straight lines which constitute the twin interfaces. As with grain boundary precipitates, the growth along the interface is favored over growth along a crystallographic plane in either adjacent crystal.

This micrograph also shows a confused state of grain boundaries. Apparently grain growth has been proceeding at the same time as precipitation since the sites of some of the old grain boundaries outlined by precipitate particles have been swept into enlarged grains and so the microstructure presents a confusion of precipitation-outlined traces of old grain boundaries and actual new existing grain boundaries.

Etchant: 20% HF, 20% HNO₃, 60% glycerin. ×250.

tion bands and slip bands can be locations at which precipitation will occur before homogeneous nucleation. Some examples are shown in Figs. 4.3, 4.4, and 4.5. Since preferences for nucleation sites do exist, it is not surprising to find that finer grain size, subgrain structures, and prior deformation lead to faster volume rates of precipitation.

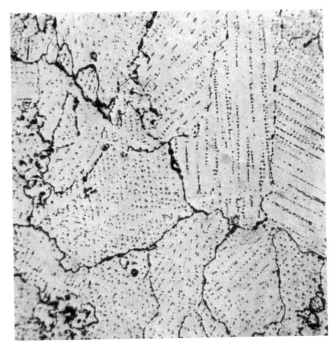

Fɪɢ. 4.4. In spite of careful handling, the lead alloy shown (0.1%Li) was slightly deformed and, on subsequent aging, the active slip planes were revealed by lines of fine particles of the precipitating intermetallic compound, PbLi. This is an example of the nucleating character of dislocation lines. An active slip band is by present theory a plane or series of adjacent parallel planes populated by concentric loops of dislocations crowding and abutting against the grain boundaries. The disposition of these dislocations is revealed by the precipitate which condensed along the dislocation lines as "dew on a spider web." In the section and planar surface preparation, the threads of precipitate are revealed as a series of points along a slip plane. The same effect can be produced by segregation of alloying elements to dislocations without actual reversion to the new crystalline form of a precipitate. In this case, the etching rate difference produces an etch pit which is often difficult to distinguish from a fine particle of precipitate.

Etchant: 15% HNO₃, 15% acetic acid, 70% glycerin. ×100.

As the degree of undercooling increases, the probability of homogeneous nucleation becomes equal to the heterogeneous case with the result that precipitation appears to begin all through a grain at the same time. The microstructure in Fig. 4.6 shows no special concentration of precipitation at grain boundaries.

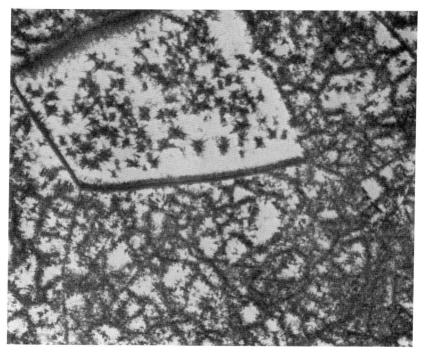

Fɪɢ. 4.5. In this micrograph, a fine, acicular precipitate of α outlines both the β phase grain boundaries and the subgrain boundaries in the β grains. The structure is one of a Ti alloy containing 11% Mo and an unusually high oxygen level of 0.55%. The prior thermal history involved a solution treatment at 1000°C in the β field followed by a quench to 700°C where the specimen was held for 5 minutes before final cooling to room temperature. The total amount of precipitate developed in this short isothermal transformation is considerably more than the same alloy exhibits at a lower oxygen level. The exaggerated rate of precipitation correlates with the existence of the extra surface area of preferred nucleation provided by subgrain boundaries which in turn correlates with the high oxygen level. Subgrain boundaries are not apparent in lower oxygen alloys having experienced identical thermal-mechanical histories. One can draw the conclusion that oxygen segregation to the subgrain bonudaries acts as a nucleating agent for precipitation of the α phase. This preference of the α-Ti phase for a high oxygen region conforms to the character of phase equilibria because in the Ti-O system the solubility of oxygen in α-Ti is nearly six times that of the β phase.

Etchant: 20% HF, 20% HNO₃, 60% glycerin. ×225.

FIG. 4.6. At very large degrees of undercooling, the grain bounary loses its preferred status as a site for nucleation of precipitates. Instead the precipitate seems to appear uniformly dispersed. Such low temperature precipitates are also usually very fine and barely resolvable. The microstructure illustrated is one of a Ti alloy containing 7% Mo which had been solution-treated in the high temperature β phase field, quenched to 500°C, and held there for 60 minutes. This represents about 315°C undercooling below the $\beta/\alpha + \beta$ transus. The very finely dispersed phase is mostly α-Titanium but X-ray diffraction indicates a substantial amount of the ω transition phase. The existence of this cannot be guessed from the microstructure. The precipitation at this temperature shows very little preference for the grain boundary.

Etchant: 20% HF, 20% HNO₃, 60% glycerin. ×500.

Morphology of Precipitate Growths

If the differences in time for heterogeneous and homogeneous nucleation are large, precipitation growths may almost all be initiated and propagated from the grain boundary. With smaller differences, the initial grain boundary precipitates grow only limited distances into the interior of the grain before they encounter equilibrium saturated matrix established by advanced stages of homogeneous nucleation. This spectrum of conditions leads to a wide variety of precipitate morphologies and arrangements. There is a reasonable state of understanding of the factors which govern the development of a precipitation structure although the appreciation of their relative importance in each instance is still imperfect.

Precipitates in the early stages of growth usually, but not always, assume a rod or thin, narrow, platelike form. It was recognized early in the history of metallography that the rods and platelets preserved specific angular relationships with each other. X-Ray diffraction studies demonstrated that each precipitate particle was a single crystallographic orientation and that the principal planes and directions of the precipitate shape corresponded to simple crystallographic descriptions. Moreover it was discovered that definite orientation relationships existed between the precipitate and the matrix solid solution whence it grew.* It has been demonstrated, for example, that the principal interfacial plane between α iron platelets in an austenite matrix corresponds to a match of the (110) plane of the body centered cubic structure of α with the (111) plane of the face centered cubic structure of γ iron and of the [111] direction in α with the [110] direction of γ. In general the matching planes and directions are close packed and likely to produce a close conformity of atomic arrangement. This is in line with expectation of the choice of a low interfacial energy arrangement. Such crystallographic matching prerequisites admit of multiple solutions so that it is possible for precipitate platelets to be other than parallel. But in each case, the angular relationships between planes and directions of precipitate and matrix are specifically those permitted by crystallography. In fact, proper measurement of these angles is a basis for experimental determination of the crystallographic habits of hitherto unknown precipitation systems.

Angular relationships between precipitate growths in any given grain or crystal originate both from accidents of nucleation and of growth. In many cases it is not possible to distinguish which origin was operative. The particular shapes of precipitation around grain boundaries provide cases in point. Figure 4.7 is presented to illustrate a remark made earlier that nucleation and growth need not always follow crystallographic discipline. Grain boundary growths whether continuous or discontinuous follow directions dictated by the boundary interface rather than by the orientations of the grains on either side. However, branching growths from these networks are frequent and these do possess obvious crystallographic discipline in form and angles of intersection. Figure 4.7(c) illustrates this. Precipitates are

* See C. S. Barrett and T. B. Massalski, "Structure of Metals" (3rd ed.). McGraw-Hill, New York, 1966.

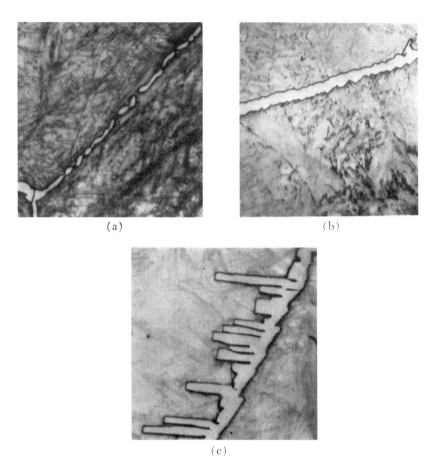

(a)

(b)

(c)

Fig. 4.7. These three micrographs illustrate the evolution of the shape character of cementite precipitating from austenite in hypereutectoid, low alloy steels. After austenitizing, these specimens were isothermally transformed (a) 800°C for 90 minutes, (b) 800°C for 2 hours, (c) 900°C for 15 minutes, and then quenched to room temperature. Cementite precipitation was confined to the grain boundaries, the remaining austenite transforming in major part to martensite on quenching.

In the beginning cementite nucleates discontinuously along austenite grain boundaries. The nuclei grow along the grain boundaries about twice as fast as into the austenite grains. While portions of austenite grain boundaries exist, growth is confined largely to lateral extension and progressive linkage until a band of cementite envelopes each austenite grain. Only then can the normal preference for growth along crystallographic planes toward the interior of each grain achieve dominance. Micrograph (b) shows the very beginning of Widmanstätten plate growths emanating from the grain boundary envelopes. Micrograph (c) shows the Widmanstätten oriented growth pattern in an advanced state of maturity. (a) and (b): Steel containing 1.2% C, 0.9% Mn. (c): Steel containing 1.48% C, 0.9% Mn. (Figures courtesy of Dr. H. W. Paxton, U. S. Steel Corp.)

Etchant: 2 gm picric acid, 1% HNO_3, 99% ethyl alcohol. $\times$700.

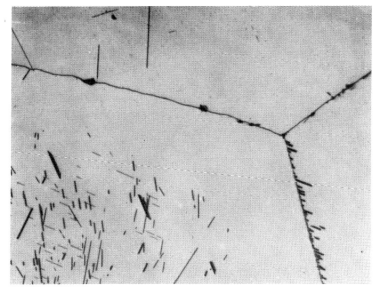

FIG. 4.8. Most grain boundary precipitation is nonspecific in shape laying like a belt around each grain of supersaturated phase. But on occasion as in this instance, the precipitate takes the form of uniform thickness, closely parallel platelets or rods. The alloy is Ti–11% Cr which has been solution-treated in the β phase field, quenched to 800°C, and held for 15 minutes before quenching to room temperature. The precipitate is the α phase which is nearly pure Ti.

Etchant: 20% HF, 20% HNO₃, 60% glycerin. ×250.

capable of branching during growth but this is less common than in dendritic growth from a melt. Not all growths from a grain boundary are preceded by the establishment of a network of precipitate as in Fig. 4.7. There are frequent occasions when a colony of parallel platelets seem simply to have nucleated at a grain boundary and grown edgewise into the interior of a grain as shown in Fig. 4.8.

Precipitates generally grow as platelets or rods and assemblies of these. They may rapidly evolve toward spheroidal shapes but this is a latter stage event. The shape assumed in growth is thought to be a compromise of the contribution of lattice strain and surface tension to the thermodynamics of the process. In certain cases, direction of maximum solute supply is a factor.

It is convenient to classify the arrangements assumed by precipitates in terms of parallel lamellae, basket weave or Widmanstätten patterns, and branching growths. Each of these is illustrated in the

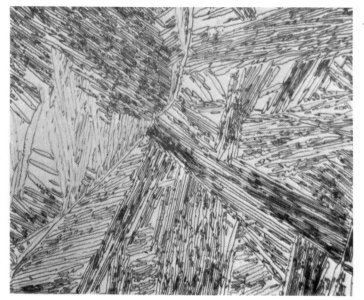

FIG. 4.9(a)

FIG. 4.9(b)

FIG. 4.9(c)

FIG. 4.9. These micrographs illustrate three general arrangements of precipitate which can be termed lamellar (a), basket weave or Widmanstätten pattern (b), and branching growth (c). All of these structures are representative of the precipitation of α-Ti from the supersaturated β phase, although each is taken from an alloy of quite different composition. (a): Ti–5% Cr–3% Al. (b): Ti–7% Mn. (c): Ti–3% Mo.

Etchant: (a), (b), and (c). 20% HF, 20% HNO₃, 60% glycerin. (a): ×250; (b): ×500; (c): ×750.

succession of micrographs in Fig. 4.9. The factors which govern the occurrence of these arrangements are far from clear and understood. It seems likely that the basket weave structure is the outcome of the growth of nuclei randomly formed throughout the body of the grain. Each nucleus grows as a platelet according to one or other of the permissible habit planes. As the planar dimensions expand they intersect each other but are impeded in growth only along the traces of intersection. The end result is that each grain becomes loosely

Fig. 4.10. The grain boundary segregated zone in this Ti alloy (10% Mo–10% V) developed from a two-stage heat treatment. The specimen after being solution-treated in the β phase field at 1000°C was air cooled. During cooling, precipitation of the α phase initiated at the grain boundaries and at selected points inside the β phase. Around each area of precipitation, a zone enriched in alloy was formed (since the α phase is almost pure Ti). On subsequent reheating to 550°C for 1 hour, α precipitation was re-initiated on a much finer scale yielding the unresolved black etching regions, but the alloy enriched zones were too close to equilibrium to permit further precipitation.

Etchant: 20% HF, 20% HNO$_3$, 60% glycerin. ×750.

separated into cells partitioned "egg crate" fashion by the precipitate platelets.

The distinction between precipitation processes at grain boundaries and in grain interiors is frequently emphasized by zones free of precipitate separating the grain boundary networks from the inside precipitation structure. This is illustrated in Fig. 4.10. The sequence of events leading to this structure is as follows. Nucleation occurred first at the grain boundaries leading to continuous or discontinuous precipitate networks delineating the original boundaries. In order to grow to resolvable size, the precipitates drew solute atoms from supersaturated solid solution in immediate surroundings. By the time the homogeneous nucleation in the interior of the grain had got underway, a thick zone of matrix around the grain boundaries had been depleted of alloy to a point where it was no longer supersaturated.

This precluded both nucleation and growth of precipitation in these regions and represents a form of solid solution microsegregation.

The evolution and growth of colonies of parallel lamellae needs some elaboration. Two types are illustrated in Figs. 4.9(a) and 4.11. In the first, the saturated interlamellar solid solution is continuous in orientation with the supersaturated grain into which the colony is growing. The second type is different in that the same interleaved saturated solid solution bands are of an orientation different from the parent grain. The interface bounding the colony establishes this. This particular structural formation has been variously termed nodular, cellular, discontinuous, or recrystallization precipitation. The latter term refers to the reorientation of the solid solution lamellae.

The origin of the lamellar precipitate colonies is probably through the following sequence of events. A single precipitate nucleus grows out from a grain boundary. A branch from a grain boundary envelope would serve the same purpose [see Fig. 4.7(c)]. The precipitate embryo will grow into the grain as a thin, narrow platelet or as a rod. This shape represents an optimum balance between strain energy and surface energy as well as providing a maximum interfacial area for most rapid absorption of solute atoms from the surrounding supersaturated matrix. With very little growth in thickness, a substantial zone on either side of the platelet becomes depleted in solute atoms to the level of equilibrium saturation.

Somehow the juncture of this concentration gradient in the immediate matrix with the surrounding uniform level of supersaturation is susceptible to nucleation of new precipitate platelets. The opinion[°] is that localized strain energy or actually plastic distortion is the key factor. Both the volume changes associated with the formation of the precipitate and the depletion of alloy from the surrounding matrix can lead to a localized stress field and preferential nucleation. Since the stress field will be symmetrical about the original platelet, the new precipitate nuclei on either side are also oriented parallel to the original platelet. With the multiple repetition of this sequence of events there is a basis for lateral growth of the colony. The criticality of the stress field intensity provides both for the spacing between the lamellae and the repetition of the spacing within narrow limits. We will carry the story one step further by supposing that if the volume change by solute depletion and the consequent local stress intensity be high enough, the side bands of saturated solid solution may choose to recrystallize or reorient for easier strain conformity. This suggests

[°] See A. H. Geisler, "Phase Transformations in Solids," pp. 432–444. Wiley, New York, 1951.

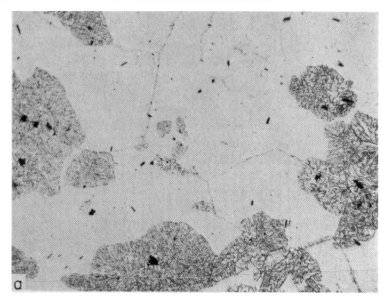

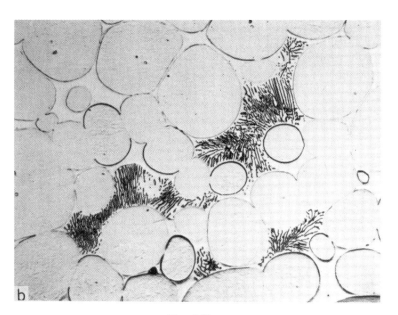

Fig. 4.11

that the relatively rare discontinuous precipitation is a special case of lamellar colony precipitation.

Temperature-Time Effects on Structure

The degree of undercooling below the solvus temperature is a dominant factor in the precipitation structure. First, with decreasing temperatures the time lag between grain boundary and grain interior nucleation gradually lessens with the consequence that the amount of grain boundary envelope precipitation decreases and finally disappears. Second, the platelet size diminishes with temperature and with it the area of lamellae colonies. At the temperature where boundary nucleation is no longer more rapid, the lamellae colonies also disappear. As illustrated in Fig. 4.12, the platelet size can diminish to beyond resolution of either shape or size by optical means. Under these circumstances the progress of precipitation can scarcely be followed by metallographic means since the only response to etching is an increase in general darkening of each grain with increasing aging time.

As mentioned earlier the platelet or rod shape is a compromise during growth of strain energy and surface energy. With the completion of precipitation, the stress fields surrounding particles can slowly relax

FIG. 4.11. While microstructure (a) gives the appearance of an intermediate state of eutectoid transformation, it is not, in fact. This U–20% Nb alloy after solution treating in the β-U phase field at 1200°C, was quenched to 750°C and held there for 90 hours. This isothermal transformation temperature is 116°C above the eutectoid temperature for the U-Nb system. The structure is an example of the type variously termed nodular, cellular, discontinuous, or recrystallization precipitation. The first three adjectives are obviously descriptive. The fourth implies that the act of depleting the matrix by precipitation has caused a reorientation or recrystallization of the matrix component left between the platelets of precipitate. This is indicated by what appears to be an interface separating the two-phase precipitation cell from the remaining supersaturated parent phase. In this instance the precipitate is a Nb solid solution isomorphous with the β-U phase.

The same interface condition can be seen in microstructure (b) more sharply in one region. In the others, the cellular growth has occupied the whole volume of previous supersaturated solid solution and so the remnants of the matrix solid solution have been completely recrystallized. This micrograph is of a W-Ni-Fe alloy (90% W–6% Ni–4% Fe). The spheroidal phase is almost pure W. The enveloping matrix is a Ni-base alloy containing appreciable amounts of W and Fe. By virtue of a prolonged heat treatment of 200 hours at 840°C, the Ni-base solid solution is induced to precipitate what is probably the Ni$_4$W phase.

Etchant: (a) 5 gm oxalic acid, 100 ml H$_2$O. ×500. (b) 20% HF, 20% HNO$_3$, 60% glycerin. ×500.

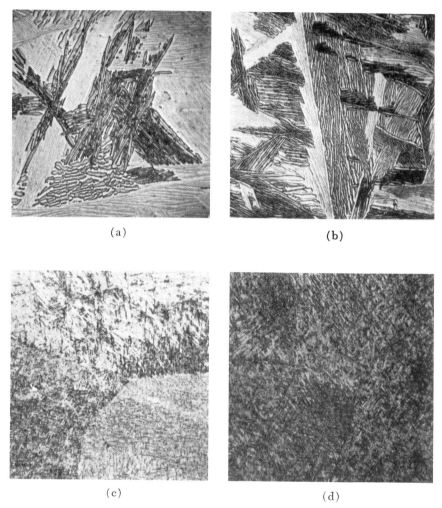

(a)

(b)

(c)

(d)

FIG. 4.12. This series of micrographs illustrates the influence of isothermal transformation temperature on the dimensions of precipitate. Note how at the lower temperature the precipitate colonies seem to disappear and the platelets shape becomes almost irresolvable. Ti alloy containing 4% Cr, 2% Mo. Solution treated at 1000°C, down quenched to a certain temperature, held for 30 minutes, and cooled. (a) 750°C, (b) 700°C, (c) 600°C, (d) 550°C. Structure: precipitation of α-Ti in a β-Ti matrix.

Etchant: (a)–(d). 20% HF, 20% HNO₃, 60% glycerin. ×365.

(b)

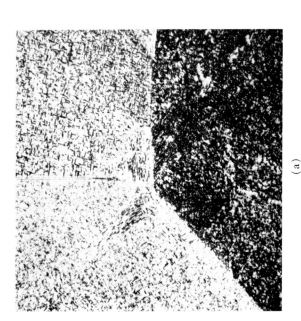

(a)

FIG. 4.13. These two companion micrographs are taken from a specimen of a Ti alloy containing 6% Cr. The alloy had been solution-treated in the high temperature β field, quenched to 700°C, and held there for 2 hours. The result was a very fine, uniform distribution of the α phase throughout the β grains and a thin, almost continuous network about the β grain boundaries. With continued heating at this temperature, the structure coarsens very slowly. The coarsened structure shown required 92 hours at temperature to develop. Various stages of growth can be seen. The short, thin platelets of original precipitate grow by both increase in length and thickness. Ultimately they lose their platelike geometry in favor of almost equiaxed blocks. The grain boundary networks also grow in thickness but with time their continuity is broken as the drive toward minimum interfacial energy gains ascendency and nodular islands of α develop.

Etchant: 20% HF, 20% HNO₃, 60% gylcerin. ×250.

Fig. 4.14 This is an example of an extremely persistent state of metastability. The structure is of an alloy of Ti with 17% O and contains an intimate mixture of the O-saturated α phase and the Ti-saturated suboxide phase, TiO, in a configuration of highly perfect alternate lamellae. The structure developed on cooling from a temperature at which TiO was the single, homogeneous phase. The α phase is therefore the precipitant. In an effort to spheroidize the structure and endow it with a more conventional two-phase appearance, the specimen was annealed at 1700°C for 1 hour but to utterly no avail. So low is the interfacial energy and so precise are the orientation relationships between the two phases that no significant driving force exists to minimize the interfacial area.

Etchant: 20% HF, 20% HNO_3, 60% glycerin. ×67 PL.

to a point where the unfavorable surface to volume shape of the particle becomes thermodynamically unstable. Toward a lower surface energy condition, the precipitate will gradually coarsen and approach a blocky shape. This is generally a slow process as indicated by annealing times for Fig. 4.13.

However, plastic deformation, cold or hot, can markedly accelerate the coarsening-spheroidization process. This is illustrated in Fig. 1.23. In certain rare instances, a lamellar precipitate arrangement utterly refuses to coarsen or change shape under any combination of temperature and time of anneal. In such cases the natural interfacial tension is extremely small and no significant driving force to coarsen and spheroidize exists. As shown in Fig. 4.14 these cases are characterized by high perfection of lamellae parallelism and periodicity of spacing. Platelet coagulation can be induced easily if the structure is capable of plastic deformation. The strain energy of distortion is sufficient to drive the process with great rapidity.

The Eutectoid Transformation

The simultaneous precipitation at low temperatures of two chemically and structurally different phases from a third solid phase stable at higher temperatures is called a eutectoid transformation. In a binary alloy system the parent phase essentially dissociates into the two product phases below some critical temperature. This is not to say that the parent phase cannot exist below the eutectoid temperature, for it can. But its condition is metastable as defined by a phase diagram. This metastability signifies a limited lifetime at any given degree of undercooling but in many instances this limited time is infinite from a practical viewpoint. The metastable phase in general has a propensity for transformation when deformed. The deformation may induce transformation toward equilibrium or to some other metastable phase such as a martensite. Continuous cooling itself may induce a martensitic transformation. Thus apparent stability, limited stability, deformation-induced or low-temperature-induced inversions are all possibilities.

In a ternary system a temperature range of coexistence of the parent and the two product phases does occur under equilibrium conditions. It is often forgotten that most alloy steels represent polycomponent systems and that there are many instances of the stable coexistence of austenite, carbides, and ferrite over broad temperature ranges.

Since the transformation process is governed by diffusion in the solid state, the high temperature phase can be undercooled and transformation rates can be sufficiently slow to permit its suppression by quenching to room temperature. Room temperature in most alloys is low enough for atomic mobilities to be essentially negligible. Even at temperatures which are high but yet below the eutectoid temperature, the metastable phase can exist untransformed for finite periods of time. The locus of stability periods as a function of temperature is of the form of a "C" which implies a maximum transformation rate at

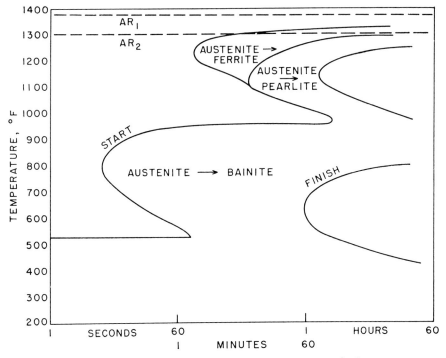

FIG. 4.15. Time-temperature-transformation diagram of a steel of composition: C, 0.42%; Ni, 1.74%; Cr, 0.81%; Mo, 0.29%; Cu, 0.10%.

some specific degree of undercooling. A single C curve governs the transformation rate by one process only. It can happen, as in the case of steels, that the formation of the two product phases at lower temperatures occur by a process different from the one at higher temperatures. In this case, the stability time-temperature zone is bounded by two superimposed C curves as shown in Fig. 4.15.

The diagram in Fig. 4.15 illustrates one of the potential sources of error in the exact determination of the eutectoid temperature for a binary alloy system. It is common practice to anneal specimens with the retained parent phase for prolonged periods at a succession of temperatures and to examine after each anneal for the appearance of a eutectoid structure. It can happen, if the anneals are of too short duration, that the recognition of the eutectoid structure pinpoints the "nose" of the C curve rather than the true equilibrium eutectoid temperature. One can check this possibility and incidentally obtain a more precise determination of the invariant temperature by reanneal-

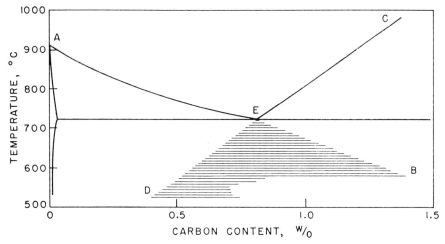

Fig. 4.16. Iron-carbon diagram showing approximate region (shaded) of simultaneous precipitation of ferrite and carbide from austenite.

ing the eutectoid structure at successively higher temperatures until the reversion to the high temperature parent phase is seen to have occurred. This is fundamentally a preferable procedure because the reversion rate is much more rapid and therefore less likely to give an incompletely transformed state.

In eutectic crystallization, the eutectic composition and eutectic temperature are narrowly limited. But it must be recognized that the eutectoid transformation is far less confined by temperature and composition. This latter point is illustrated in Fig. 4.16. Lines AB and CD represent the loci of temperatures below which the high temperature phase is unstable with respect to one or other of the product phases. The area DEB represents the temperature-composition range in which the high temperature, parent phase can reject both phases simultaneously. By definition, this coprecipitation must be within the context of a eutectoid transformation. The eutectoid structure itself, therefore, may contain wide variations in the volumetric proportions of the two product phases. Thus, for example, in plain carbon steels, the eutectoid of ferrite and carbide (pearlite) can be constrained to occupy the whole structure over the carbon range 0.4–1.2% although the eutectoid composition as dictated by the Fe-C phase diagram is about 0.8% C. This particular point of discussion applies to circumstances where the parent phase enjoys a wide miscibility range such that the area subtended by the lines AEB and CED (Fig. 4.16) is significant. There are, however, numerous intermediate phases which dissociate at low temperatures by a eutectoid transformation but possess an almost unique stoichiometry.

The occurrence of a broad composition range of the eutectoid structure hinges on the existence of almost equal rates of precipitation of the two product phases or at least a mutually stimulated nucleation. This does not always happen and, in many steels and titanium alloys, the extensive formation of hypoeutectoid precipitants precedes the nucleation of the eutectoid structure. Under such conditions the eutectoid structure is likely to occur at the ideal composition by virtue of alloy enrichment of the parent phase.

The eutectoid transformation product is most commonly recognized in the form of colonies of what appear to be alternating and almost parallel strips of two-phase species. Actually this structure is one of almost parallel, discontinuous plates of one-phase species immersed in a continuum of the other phase species. In this respect the lamellar eutectoid structure and the lamellar eutectic structure are identical in nature. In spite of a proliferation of other theories* the origin and growth of lamellar eutectoid structures is thought now to closely parallel the lamellar eutectic. One major point of difference is that temperature gradients are not a factor in eutectoid nucleation and growth processes. The transformation can proceed under isothermal conditions. Nucleation occurs primarily at grain boundaries of the parent phase or on to the surfaces of certain proeutectoid precipitates. The continuity of proeutectoid phases with their own kind in the eutectoid complex is strong metallographic evidence for identifying preferential nucleation sites as illustrated in Fig. 4.17.

The eutectoid transformation of austenite in iron-carbon alloys is both technically and scientifically the most interesting case. In pearlite, ferrite is the continuous and cementite, the discontinuous, plate-shaped phase. We may argue that ferrite nucleates at austenite grain boundaries and grows along the boundary and into one grain where a preferred ferrite-austenite orientation exists. Cementite nucleates at periodic points along the ferrite-carbon enriched austenite interface. The fact that cementite grows as roughly parallel plates demonstrates that the nuclei form on the ferrite and possess rigorous orientation relationships with the latter because of lattice conformity restrictions. As in the eutectic structure, the discontinuous cementite phase grows only a limited distance before the advancing edge is closed off by the

* For detailed discussion see R. F. Mehl and W. C. Hagel, *Progr. Metal Phys.* **6.** 74–134 (1956).

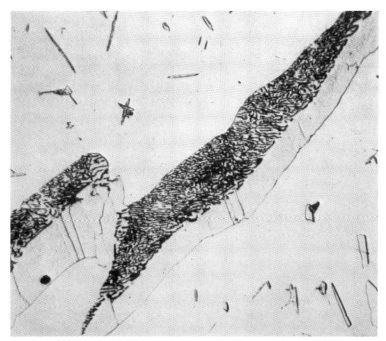

FIG. 4.17. The δ phase, Cu₃₁ Sn₈, is the γ brass analogy in the Cu-Sn system. For many years it was thought to be stable to room temperature. X-ray diffraction studies on powdered samples annealed for several days at low temperatures were the first indications that the phase undergoes a eutectoid transformation below 350°C. The microstructure shown is the first metallographic evidence of this. The specimen contained 32.6% Sn and as-quenched from 400°C was predominantly the high temperature, δ phase with a minority distribution of the ε phase, Cu₃Sn as platelets arranged in a Widmanstätten pattern. After annealing for 63 days at 300°C the eutectoid decomposition δ → α + ε has just begun. This micrograph is particularly interesting because it shows that the proeuctetoid ε phase nucleates the eutectoid transformation. The lamellar growths on one side of each of the large platelets of ε are so oriented that the ε lamellae of the eutectoid structure are continuous with the proeutectoid ε. There is no interface between them. In one zone, the identical orientations of the two major proeutectoid platelets of ε are demonstrated by the continuity of the ε lamellae of the eutectoid with both platelets. The sluggishness in transformation of massive specimens compared to high specific powders can be appreciated from the comparative rates in this instance. The X-ray diffraction studies indicated that the alloy powders were almost completely transformed in 5–7 days at 300°C, whereas the lump specimen, of which the micrograph is exemplary, had barely begun to transform in 63 days. [See C. C. Wang and M. Hansen, *Trans. AIME* **191**, 1212 (1951).]

Etchant: 5 gm FeCl₃, 100 ml H₂O. ×250.

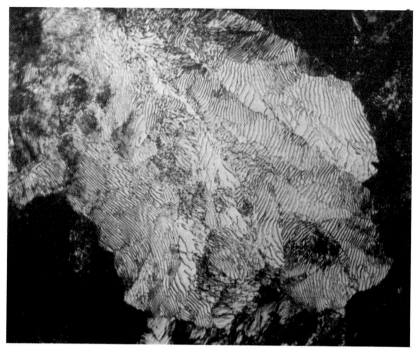

Fɪɢ. 4.18. This typical micrograph of pearlite in steel illustrates several points in the character of the lamellar eutectoid. The platelets are parallel to each other on an average over their whole length. The cementite plates do not follow a specific crystallographic direction rigorously for they have substantial curvature. Platelets do not extend over the full dimension of the colony. There are numerous instances apparent where the growth of a plate is cut off by its surrounding neighbors, which is further evidence for the nonrigorously crystallographic growth character of the plates. The obvious variation in cementite plate thickness and spacing indicates how the major axes of the eutectoid colonies are inclined to each other in three-dimensional space.

Etchant: 2% HNO₃, 98% ethyl alcohol. ×900.

surrounding ferrite matrix phase. Note that the thicker the plate the more difficult it is to effect edge closure and so thick plates are also long plates.

The cementite plates in Fig. 4.18 show considerable waviness although their average direction is clearly defined. The waviness of the plates indicates that no rigorous orientation relationship exists between cementite and the parent austenite. As may be seen by comparing Fig. 4.18 with Fig. 4.19, the spacing between cementite lamellae

Fig. 4.19. This micrograph of pearlite is to be compared with Fig. 4.18. There are certain notable differences in the initial conditions and final structure. This micrograph is taken from an SAE 1040 steel, so there is a substantial volume of ferrite associated with the pearlite. The distribution of the pearlite regions illustrates the pattern of growth. At first glance, the pearlite regions seem to form an intercrystalline network. This is true in a sense but not by primary choice of growth directions and locations as in precipitation from singly supersaturated solid solution. In this particular view, the section cuts through a sheaf of coarse rods and narrow plates of proeutectoid ferrite. The austenite between the rods, when sufficiently enriched with carbon, transformed to pearlite. The original austenite grain from which these events derived is probably several times the size of this area in view. The transformation to pearlite took place at a significantly lower undercooling temperature than in Fig. 4.18 as indicated by the thinner and shorter average platelet size of cementite in the pearlite colonies. Etchant: 2% HNO_3, 98% ethyl alcohol. $\times 1000$.

decreases with lower transformation temperatures. This primarily reflects the increased rate of carbide nucleation with increased undercooling, i.e., more nuclei of carbide from per unit of time and per unit of ferrite-austenite interface at lower temperatures.

If the second phase species of the eutectoid cannot be nucleated

(a)

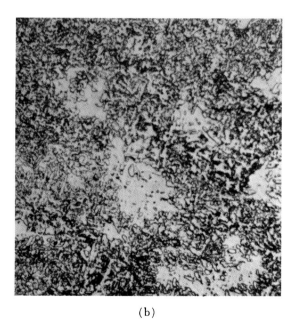

(b)

FIG. 4.20

by the first phase species, then alloy enrichment proceeds in the zone of parent phase immediately adjacent to the transformation interface until nucleation is induced. These nuclei will not be parallel to each other and a nonlamellar eutectoid structure will result. Mehl and Hagel remark upon such instances (p. 95 of footnote reference).

Eutectoid colonies will nucleate at grain boundaries preferentially if this is not forestalled as is more commonly the case by prior precipitation of proeutectoid constituents. Each austenite grain may produce several colonies whose principal directions differ considerably. With the complete transformation of austenite to proeutectoid and eutectoid constituents, the original orientation identity of the austenite grain has disappeared. On reaustenitizing, each pearlite colony has several optional austenite grain orientations to choose, with the result that different choices are made and the new austenite grain size of necessity is smaller than the original. This grain refining capability in an alloy is unique to the eutectoid type transformation. Simple precipitation and re-solution treating cannot produce this because the parent phase and its orientation identity are preserved.

The lamellar structure is not of itself the fingerprint identification of a eutectoid transformation for the cellular form of precipitation from a supersaturated solid solution has all of the morphological features of a eutectoid (see Fig. 4.11). They can be distinguished in one of two ways. In X-ray diffraction identification of the phases, the

---

FIG. 4 20. These two companion micrographs illustrate an incompleted stage and the completed stage of the eutectoid transformation in a Ti alloy containing 9% Cr. This is a hypoeutectoid composition and the eutectoid transformation itself is very sluggish. The alloy was solution-treated at 1000°C and quenched to 635°C and held for various periods of time. The proeutectoid α precipitates rapidly from the high temperature β phase and, in a matter of a few hours, the structure is fully populated with the Widmanstätten arrangement of platelets. For some time thereafter the structure appears stable. The first micrograph (a) shows the beginning of the eutectoid transformation in the matrix β phase after hours at temperature. This is an example of a nonlamellar eutectoid. The growth rate is literally so slow that the kinetics gain no advantage in a lamellar form. One might also say that the lamellae coagulate at the same rate as they form. The α phase portion of the eutectoid phase mixture ($\alpha + TiCr_2$) becomes an ill-defined background and as the eutectoid transformation invests the β phase residue between the proeutectoid α plates, the original Widmanstätten pattern disappears as in the micrograph (b) representing 10 days of isothermal heating. This is really an etching effect. The $\alpha/TiCr_2$ interfaces etch so rapidly in comparison with the α/α interfaces that the microstructure etches completely black before the α grains can be resolved.

Etchant: (a) and (b). 20% HF, 20% HNO$_3$, 60% glycerin. $\times 675$.

parent phase remains one of the constituents in the cellular precipitation structure whereas, in the eutectoid structure, the parent phase does not exist unless the transformation state is only partial and preserved to room temperature. Alternatively, variation in alloy composition can permit distinction because only in the eutectoid system is the re-solution temperature independent of composition. Moreover, the eutectoid suddenly disappears with increasing temperature whereas the simple precipitate redissolves progressively.

Not only is the lamellar structure not the exclusive result of a eutectoid transformation but eutectoid transformations do not invariably produce lamellar type structures. Figure 4.20 illustrates an intermediate state in the eutectoid transformation of a hypoeutectoid alloy in the titanium-chromium system. The compound $TiCr_2$ is clearly growing in a nodular form as a dispersion between platelets of proeutectoid α. When the transformation is complete, the original structure of the proeutectoid constituent has disappeared. This is because the intermetallic compound nucleated at the α/β interface leading to simple growth of the α phase at the expense of β. Since no new α orientations were generated, the resulting structure is simply a dispersion of the compound phase in an expanded version of the original proeutectoid α phase structure. The α/α boundaries are not resolvable because of the etching dominance of the $\alpha/TiCr_2$ interfaces. A lamellar eutectoid structure can be produced in titanium-chromium alloys but only in hypereutectoid compositions and at temperatures near the eutectoid invariancy.

In any one alloy or alloy system, the process of eutectoid transformation can change. The alloyed carbon steels and cast irons provide examples of this in the change in transformation from pearlite to bainite. Bainite, as illustrated in Fig. 4.21, is the same two phases formed by a different process. Its individuality is emphasized by the superposition of a separate C curve of nucleation as shown diagrammatically in Fig. 4.15. As it is presently believed, bainite represents a process whereby ferrite supersaturated with carbon is nucleated from austenite. As the supersaturated ferrite grows it rejects carbides in a finely dispersed form sometimes at the austenite-ferrite interface but not invariably. Moreover, the carbide phase need not be cementite but may be of some metastable or transition form with limited stability in temperature-time space. The low temperature range of dominance of the bainitic process leads to dispersed phase structures which are generally beyond the resolution capability of optical metallography.

FIG. 4.21. This steel has the composition: C, 0.76%; Mn, 0.52%; Cr, 0.25%. Its prior thermal history was as follows: austenitized, quenched to 300°C, held for 6 minutes, water quenched. During the time at 300°C, bainite formed isothermally. The remaining austenite transformed to martensite on quenching. The bainite is quite dark etching which makes it difficult to resolve the martensite. The individual regions of bainite while possessing some of the acicular characteristics of martensite are in general more ragged. The actual structure of bainite cannot be resolved at optical magnifications.

Etchant: 2% HNO₃, 98% ethyl alcohol. ×800.

For this reason, it has been very difficult to study bainite formation as a process in its relation to time, composition, and prior structural parameters. Bainite is a darker etching phase than untempered or lightly tempered martensite. Also bainite regions have a morphology usually distinguishable from martensite. However, at low carbon contents tempered bainite may be difficult to distinguish from martensite by normal optical appearance.

It is relatively unusual to find a structure which is wholly composed of a eutectoid structure. More commonly the appearance of eutectoid is preceded by the independent precipitation of a pro-

(a)

(b)

FIG. 4.22

eutectoid constituent. The amount and distribution of the proeutectoid constituent may have a profound influence on the mechanical properties of the structure. Perhaps the most damaging to toughness is the distribution of the proeutectoid phase as a band around each grain of the former parent phase. When the nucleation rate is low, the precipitation may be confined to relatively few sites leading to a large nonoriented growth such as the illustration of "blocky" ferrite in Fig. 4.22(a). The appearance of blocky ferrite can be synthesized by a subcritical solution-treating or austenitizing treatment but the structure in this case is a residue from the prior structure rather than the product of a new transformation [Fig. 4.22(b)]. In the dilute, hypoeutectoid alloys, the proeutectoid phase (terminal solid solution) is the major constituent, and the eutectoid appears only in interstices of the interleaving plates of proeutectoid phase (see Fig. 4.23) or in isolated pockets. The latter will occur when the proeutectoid precipitation is not heavily oriented and can develop an equiaxed, polycrystalline character.

The morphology of proeutectoid constituents can be related to their position in the temperature-composition space of a phase diagram.

FIG. 4.22. (a). This is the structure of a specimen of AISI 4340 steel which, having been austenitized at 1550°F, was thereupon quenched to 1200°F, then held at that temperature for 5 minutes prior to water quenching. During the interrupted quench at 1200°F, proeutectoid ferrite nucleated at many points and grew in a nonoriented fashion to provide a pattern of "blocky" ferrite. The remaining austenite transformed to martensite on subsequent quenching. The blocky ferrite is more easily resolvable in the quenched or lightly tempered form. As the carbides become resolvable at the higher tempering temperatures, the zones of ferrite become less distinct.

Etchant: 2% HNO_3, 98% ethyl alcohol. ×500.

(b) Large, blocky areas of a proeutectoid phase can originate either as a natural consequence of precipitation and growth or as the residue of a prior structure not completely eradicated by the latest heat treatment. Blocky ferrite of the latter origin is shown in this micrograph. The steel is AISI 8620H and in its original condition was annealed so that small areas of pearlite existed in a polycrystalline matrix of ferrite. On re-heat treatment for hardening, complete re-austenitizing was not achieved so that at the time of quenching isolated areas of undissolved ferrite remained in the austenite and these were retained on quenching and subsequent tempering. The quench itself was not very severe for the austenite partially transformed to bainite before reaching the Ms temperature. The structure reveals a mixture of tempered bainite and martensite as a matrix for the islands of undissolved, blocky ferrite. Nominal composition of AISI 8620H steel: 0.18–0.23% C, 0.70–0.90 Mn, 0.40–0.70 Ni, 0.40–0.60 Cr, 0.15–0.25 V.

Etchant: 2% HNO_3, 98% ethyl alcohol. ×150.

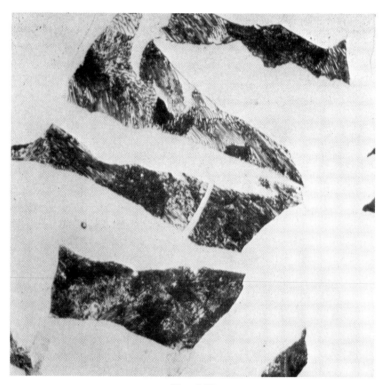

Fig. 4.23

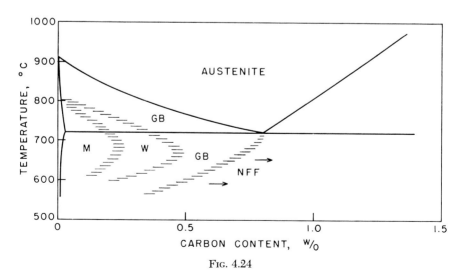

Fig. 4.24

This has been done for proeutectoid ferrite in plain carbon steels as illustrated in Fig. 4.24.

The Peritectoid Transformation

The term "peritectoid" is applied to the solid state transformation wherein a phase mixture above a critical temperature is replaced by a single new phase of a specific composition below that critical temperature. On either side of the specific composition, excess of one or other of the original high temperature phases may persist. Since one of the parent phases is richer and one poorer in alloy content than the peritectoid product phase, it is reasonable to expect the latter to nucleate at the interface between the two parent phases where the adjustment to a new chemistry can most expeditiously occur.

When the changes in chemistry are large and the transformation temperature low (so that diffusion rates are correspondingly slow), the peritectoid transformation may be easily suppressible or, at best, very difficult to bring to completion. One such example of a sluggish peritectoid transformation is illustrated in Fig. 4.25. As in the case of eutectoid transformations, the interplay of the temperature dependencies of nucleation and growth make for the most rapid rate of transformation at some degree of undercooling below the equilibrium temperature. A TTT diagram for a peritectoid transformation has the appearance of a typical "C" shaped curve.*

As may often be the case, a given level of undercooling for a particular peritectoid transformation may also involve metastability with respect to other phases. A case in point is the region of the copper-

* See R. D. Reiswig and D. J. Mack, *Trans. AIME* **215**, 301–307 (1959).

Fig. 4.23. In a AISI 1022 steel, ferrite is usually the majority phase. In the primary decomposition of austenite, ferrite grows in coarse plates across the austenite grains. Many of the interfaces between intersecting plates of ferrite can disappear quickly, reorganizing to approximately equiaxed grains. Some time is required and so approximately equiaxed grains shows primarily in normalized and annealed structures. However, the zones of trapped austenite of eutectoid composition preserve their original shape in the form of the strip-shaped zones of pearlite seen in this micrograph. Furthermore, elongated zones of ferrite trapped between pearlite zones must hold their shape.
Etchant: 2% HNO_3, 98% ethyl alcohol. ×500.

Fig. 4.24. Existence diagram for various structural forms of proeutectoid ferrite. M: massive ferrite; W: Widmanstätten network; GB: grain boundary ferrite; NFF: no free ferrite. [After R. F. Mehl and C. A. Dube, "Phase Transformations in Solids," p. 545. Wiley, New York, 1951.]

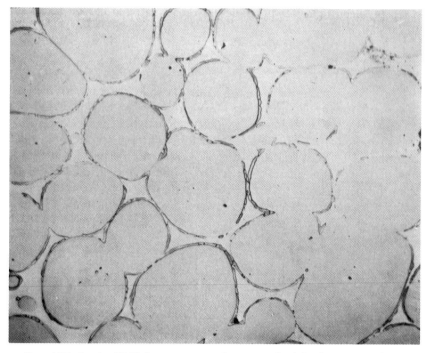

FIG. 4.25. In the W-Ni binary system the terminal solid solutions transform to the Ni_4W phase (43 w/o W) below 970°C. This peritectoid transformation is very sluggish. The alloy illustrated contains 90% W and 10% Ni and was annealed at 954°C for 200 hours to produce the rims of Ni_4W. Examination of the structure of the rim shows that the product phase was nucleated at points along the W/Ni interface and that growth was faster along the interface than away from it. Moreover, the existence of interfaces in the Ni_4W rims running perpendicular to the original Ni/W interface demonstrate that some variation in orientation of the product phase nuclei is possible.

Etchant: 33% NH_4OH, 33% H_2O_2, 34% H_2O. ×450.

antimony phase diagram between 10 and 35 a/o antimony shown in Fig. 4.26. The transformation $\alpha + \beta \to \gamma$ is easily suppressed by quenching from above 488°C. In such an event, the α phase is retained, but the β phase transforms insuppressibly to a metastable form designated β'. If one quenches a specimen containing, for example, 16 a/o antimony from 500°C to 450°C, the $(\alpha + \beta)$ phase mixture is doubly metastable with respect to the formation of γ and of δ. Which of these forms first depends on their respective nucleation rates, and these cannot be anticipated. As it happens, the δ phase nucleates first.

Because of such complications, the recognition of the existence of

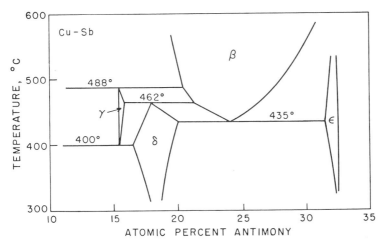

FIG. 4.26. Partial phase diagram of the Cu-Sb system.

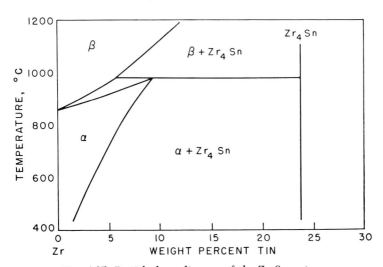

FIG. 4.27. Partial phase diagram of the Zn-Sn system.

a peritectoid transformation may be impossible from one microstructure. Even the recognition of the characteristic rim of product phase at the parent phases' interface is not enough since this could also be the consequence of a suppressed peritectic crystallization from the melt. Moreover, with large degrees of undercooling and without the interference of side transformations, the formation of the product phase may not form a continuous rim but nucleate and grow as ran-

(a)

(b)

FIG. 4.28 These two micrographs show the structure of a Zr–7.9 w/o Sn alloy annealed at (a) 1006°C for 45 hours and (b) 957°C for 70 hours which represents temperatures above and below the peritectoid temperature for the transformation β-Zr + Zr₃Sn → α-Zr. The β-Zr phase transforms insuppressibly on quenching to the α-Zr phase yielding a structure which is finely interleaved bands of the quench transformation product. The existence of the β-Zr phase at the high temperature is therefore identified by the quench transformation structure. A few particles of Zr₃Sn are seen in the transformed β-Zr structure. These can be distinguished from α-Zr by their hardness. The micrograph of the alloy annealed at 957°C shows primarily α-Zr with small amounts of transformed β-Zr. The polygonal grain boundary structure of the α-Zr demonstrates that this phase existed as shown at the temperature of annealing. The conversion from a predominantly α-Zr structure to a predominantly β-Zr structure over such a narrow temperature range at well above the α/β inversion for pure Zr (880°C) indicates the peritectoid nature of the transformation.

Etchant: (a) and (b). 20% HF, 20% HNO₃, 60% glycerin. ×150.

dom blocks. Just as the lamellar colony is not the exclusive configuration for a eutectoid transformation, so also the interface rim of product phase is not the exclusive trademark of the peritectoid transformation. The resolution of such questions requires the identification of the phases either by their etching character or by X-ray diffraction analysis. Frequently, this can be done by metallography alone. Take the case of the zirconium-rich end of the zirconium-tin phase diagram for example. As shown in Fig. 4.27, at 980°C the 9 w/o tin alloy transforms from a phase mixture of β-Zr $+$ Zr$_4$Sn to α-Zr. The structures of a 7.9% tin alloy annealed above and below the invariency temperature are shown in Fig. 4.28 to illustrate the profound change in structure over a narrow temperature range.

Martensitic Transformation

Martensites have far fewer specific attributes than were once thought. The transformation product has the same composition as its parent phase, but this can also be attained by nucleation and growth processes. Martensites are only sometimes metastable in their crystalline form and supersaturated with respect to a phase mixture. The structure of a martensite is invariably different from its parent phase but is often identical with one of the equilibrium phases which it supplants. For example, the martensite formed by quenching γ-FeNi alloys is body centered cubic just as the α iron which forms by prolonged annealing. The martensite, of course, contains the nickel content of the original austenite while the α iron at equilibrium is a much more dilute solid solution. In contrast, one of the martensites formed by quenching β-CuAl is hexagonal close packed which is different from either of the equilibrium phases produced by the alternative eutectoid transformation. This underlines another characteristic of the martensite transformation—namely, that the transformation product can change its crystallographic character within a narrow range of composition.

Since a martensite has the same composition as its parent phase, the transformation does not need to involve diffusion of more than one or two atomic distances. This permits the martensite transformation to occur very rapidly at speeds of the order of the velocity of elastic waves. But it is known that in certain instances martensitic transformations can proceed isothermally over prolonged periods of time. These time periods are mostly of inactivity—the actual time of growth of an individual martensite particle still being very short. The

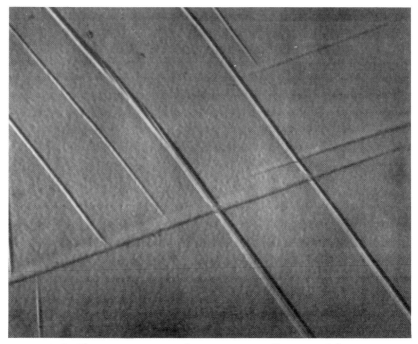

Fig. 4.29. The martensitic transformation is usually initiated by temperature depression, sometimes by cold work. This example is unusual in that martensite plates popped out of the surface during etching in an HF-HNO₃ solution. The alloy in Ti–13% Mo which had been solution-treated at 1000°F and water quenched to yield a retained bcc single phase, β-Ti structure. The appearance of the martensitic α′ needles occurred during etching. This is not a case of resolution by differential chemical attack on an existing structure. These needles actually formed during etching. The surface relief effect is brought out by the use of oblique illumination. One can see the same phenomenon with thermal martensite by the use of a hot stage microscope.

Unetched. ×675, oblique illumination.

discontinuous character of growth is one of the distinctive attributes of a martensitic process.

While undercooling is the primary factor in the initiation of the martensitic transformation, both plastic and elastic strains can exert important influence. Many martensites can be induced to form by cold deformation of the parent phase at temperatures at which the latter is apparently stable. Frequently deformation martensites are crystallographically dissimilar to thermal martensites produced from the same parent phase.

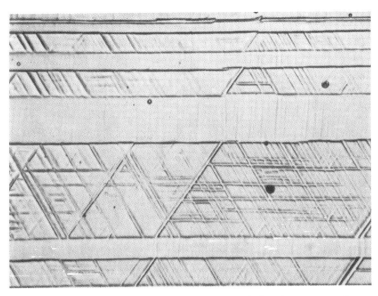

FIG. 4.30. This micrograph illustrates the banded substructure of parallel martensite plates in a Co–25% Cr–10% Mo alloy. The specimen was annealed at 1300°C for sufficient time to coarsen the grain size of the β-Co (fcc) solid solution stable at that temperature. It was then water quenched causing the inversion of β-Co to α-Co (hcp) by a martensitic process. The substructure of the martensite plates are probably mechanical twins.

    Etchant: 20% HCl, 80% acetic acid. ×1000.

The formation of martensites generally occurs below the strain relaxation temperature of the parent phase. In order to minimize the strain energy of the new system, the transformation must proceed in a manner which permits maximum conformity between the parent and product phases. To accomplish this, martensites assume a lath or needle shape with the major axis coinciding with some rational crystallographic direction in the parent phase. Martensite plates such as shown in Fig. 4.29 are very much like mechanical twins in appearance. In fact, martensite formation and mechanical twinning are often associated (see Fig. 4.30). The ribbed substructure often seen in large plates of martensite in high carbon steel have been identified as twins. It seems likely that considerable similarity exists between the processes of martensite and mechanical twin formation.

    Martensites nucleate below some critical temperature generally designated the "Ms" temperature. From strain energy considerations, the extension of a nucleus can proceed much faster than thickening.

Fig. 4.31

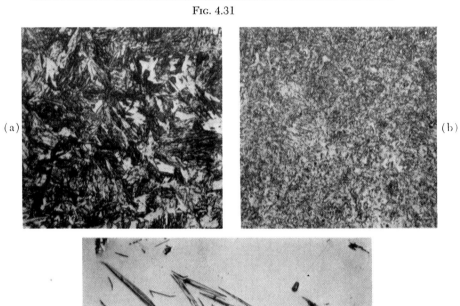

(a)

(b)

(c)

Fig. 4.32

168

Nucleation starts inside a grain and propagates in two opposite directions to the grain boundaries. Lateral growth is thus maximum in the vicinity of initiation and least at the points of final stoppage. This gives martensite needles the pointed appearance frequently observed. Growth is stopped by grain boundaries and often (but not always) by existing needles in the path of growth. This leads to a sequence of first, second, and third generation of martensite needles; each generation hemmed in by the previous ones as shown in Fig. 4.31.

Because a martensite transformation proceeds in discrete units of platelet formation, the temperature of first detection of martensite, M_s, should be dependent on grain size. The finer the grain size, the

FIG. 4.31. The first plates of martensite traverse the distance between grain boundaries. The second generation traverse the distance between plates of the first generation, and so on. This micrograph of a Co–15% Mo–0.4% Ce alloy shows at least three generations of martensite growths. The alloy was annealed at 925°C for 24 hours to establish an equiaxed grain structure of β-Co (fcc). The 15% Mo is just beyond the solid solubility limit at this temperature and so the grain boundaries of β-Co contain particles of the intermediate phase $CoMo_2$. The reversion of β-Co to α-Co (hcp) is very sluggish at the phase boundaries. However, at some considerable degree of undercooling, the transformation initiates abruptly by a martensitic process. This is an instance where an equilibrium phase is produced by a martensitic transformation. The structure illustrated developed as the result of air cooling a small sample.
Etchant: 20% HCl, 80% acetic acid. ×1000.

FIG. 4.32. This trio of micrographs is presented to illustrate the influence of carbon content on the appearance of martensite in steels—(a) 0.18% C, (b) 0.90% C, (c) 1.2% C. While the morphology of the individual martensite particles is not resolvable in either the 0.18 or 0.90% C steels, there are clear differences in general pattern of appearance. There is much more of a mottled appearance to the low carbon martensites. They are nearly indistinguishable from bainitic transformation products. There is some capacity to resolve individual plates. The structure of martensite in the 0.90% C steel is nearly amorphous. Note the carbide nodules randomly dispersed which were not taken into solution in the austenite prior to the quench. This verifies that the steel is hypereutectoid in nature.

The morphology of martensite can only be observed in the presence of a majority of retained austenite. This condition in (c) was encouraged by the presence of the high carbon content and 3% Ni. Martensite plates have a definite featherlike substructure and lenticular shape. The center midrib along the length of each plate is typical of high carbon martensites. Notice that the interfaces between martensite and austenite are periodically serrated by secondary martensite subgrowths. The arrowhead and zigzag pattern of adjacent plates suggests that their genesis has been a sequence of events by which the growth of a new plate was triggered by a neighbor.
Etchant: 2% HNO_3, 98% ethyl alcohol. (a) and (b) ×630, (c) ×415.

Fig. 4.33. This could be the structure of martensite in retained austenite in a hypereutectoid carbon steel, but it is not. The morphology of phases is not enough to characterize the transformation and crystallization history of a specimen. Actually this is the structure of a Cu–22 w/o Y alloy which after freezing was annealed at 700°C for 350 hours. At this composition, primary crystals of YCu_4 form as long needles. The dominant chill direction forced growth of the primary needles to conform in the main to this direction so that the needles are inclined to each other at acute angles. Before freezing is complete, the remaining liquid interacts with YCu_4 to bring about the peritectic crystallization of YCu_6. The long anneal served to eliminate the vestiges of nonequilibrium eutectic of YCu_6 and Cu residual in the cast structure as a result of the incomplete peritectic process.

Etchant: 2% HNO_3, 98% ethyl alcohol. ×450.

lower should be the Ms temperature. This is in contradistinction to nucleation and growth processes whose initiation is stimulated by finer parent phase grain size.

The angles of intersection of adjacent individual needles are often very acute giving the characteristic "arrowhead" pattern shown in Fig. 4.32 (c). This bespeaks the high indice crystallographic directions of the parent phase along which martensites prefer to grow. Such

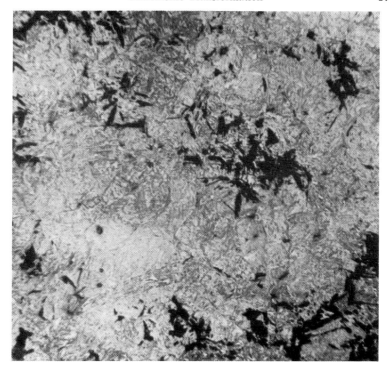

FIG. 4.34. The act of tempering martensite in carbon steels is recognizable metallographically by the more rapid attack of etching solutions. For the same time of immersion tempered martensite etches much darker than freshly quenched martensite. This fact is used in a well-known method for determining the Ms temperature (temperature below which martensite forms) of plain carbon and alloy-carbon steels. The procedure is as follows: Specimens of the steel are quenched from the temperature range of austenite into a bath held somewhere near the Ms temperature. The hold time in this bath is only sufficient to assure uniformity of temperature and then the specimen is plunged into a second bath at a higher temperature, usually about 1000°F, held for no more than a few minutes, and then water quenched. If the first bath temperature is below the Ms, austenite will transform to martensite in an amount proportionate to the degree of undercooling below the Ms. Upquenching tempers this martensite but the hold time must be short enough to prevent transformation of the remaining austenite to pearlite or upper bainite. The final quench produces substantial conversion of the remaining austenite to martensite. We have thus two generations of martensite, one tempered and one untempered.

An example of this is shown in the accompanying micrograph. The steel is AISI 4350 which has been quenched from the austenite temperature range to 400°F, held 10 seconds, upquenched to 1000°F, held for 1 minute, and water quenched. The acicular black areas represent the small amount of martensite (now tempered) formed in the initial quench. The lighter background is untempered martensite formed in the final quench. The small amount of tempered martensite indicates that the Ms temperature is only slightly above 400°F.

Etchant: 2% HNO_3, 98% ethyl alcohol. ×250.

specific shape and distribution characteristics are observable only when the transformation is partially completed. When the martensitic transformation consumes the parent phase, the structural specifics are essentially lost in the lack of contrast as may be seen in Fig. 4.32(a,b).

Transformations to martensite are accompanied by large anisotropic distortions. This is most effectively demonstrated in those cases where it is possible to retain the undisturbed parent phase to room temperature. Such a specimen may be polished and installed in a fixture under a metallurgical microscope which permits subzero cooling. As the temperature falls below the Ms, the formation of martensite will be accompanied by the appearance of acicular shapes standing in relief. On warming up again, if the martensitic transformation is reversible—as are most—the surface roughening will disappear and the polished surface will assume its previous smoothness. This thermal generation of surface disturbances is one of the valid identifications of a martensitic process.

There are few metallographic attributes to martensites which are absolutely unique to this type of transformation product. The lath-shaped martensites look very much like mechanical twins. Sometimes a distinction can be made with polarized light if only one of the crystallographic forms is optically active. Even the arrowhead arrangement can be the product of an entirely different metallurgical process as illustrated in Fig. 4.33. The striated substructure or a center midrib with striations emanating from either side are probably peculiar only to martensites (see Fig. 4.32c). But not all martensites possess this character.

The martensite plates or needles are often supersaturated solid solutions and so, on reheating to temperatures below the stable phase field of the parent phase, equilibration must be approached by precipitation of a solute-richer or solute-leaner phase to relieve the state of supersaturation. Such is the nature of tempering. It may lead either to hardening or softening.

In either case, the clear morphology of the martensitic phases can disappear in the blur of a fine particle distribution as shown in Fig. 4.34. Not all martensites are supersaturated solid solutions as in the case of the cobalt alloys illustrated. In such event, reheating below the phase boundaries can have the effect only of slow readjustment of interfaces to a more stable arrangement of equiaxed form. This might be likened to recrystallization of a plastically deformed metal.

The allotropic transformation of a pure metal is a special case of

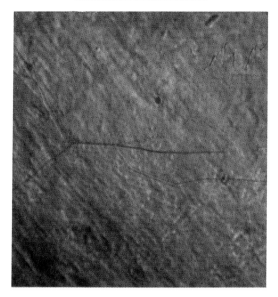

FIG. 4.35. This is a micrograph of iron taken on a hot stage microscope at 935°C. At room temperature, standard etching revealed a three-grain assembly with the sharply defined grain boundaries shown. The later grain boundary movements and the generation of new grains of austenite are revealed by selective evaporation of metal from the new locations of grain boundary and are much more subtle in appearance. One can see in this micrograph evidence of grain boundary migration before the $\alpha \rightarrow \gamma$ inversion. The $\alpha \rightarrow \gamma$ inversion itself produced a new generation of grains characterized by a very serrated system of grain boundaries typical of the martensitic process by which the transformation occurred.

Thermally etched. ×350.

the martensitic process. The acicular morphology is not often apparent. A structure results which is most commonly recognizable by the serrated appearance of the grain boundaries. A case in point is illustrated in Fig. 4.35. With more prolonged heating, these new grain boundaries will straighten out, and evidence of the prior transformation will disappear.

When the two allotropic modifications of a metal have substantial solid solubilities for certain alloying elements, the two homogeneous phase fields are separated by a phase field in which both allotropes can exist in physical mixture (but of different composition). In this case, the inversion from the high to the low temperature modification

(a)

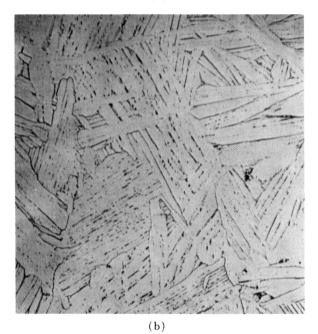

(b)

Fig. 4.36

can proceed either by nucleation and growth or by a martensitic process. The choice of process taken is usually distinguishable by metallography as shown in Fig. 4.36, but the interpretation is only possible in the light of full knowledge of the thermal history and of the phase relationships operative.

Fɪɢ. 4.36. These two micrographs illustrate how the transformation from all β phase to all α phase in a Ti–10 w/o Ta alloy can occur by either a nucleation and growth or by a martensitic process. In both instances, specimens were solution treated in the β field at 1025°C. In the case of (a), the specimen was quenched to a bath at 600°C, held there for 30 minutes, and then water quenched. In the case of (b), the specimen was water quenched directly from 1025°C. Both of these structures despite their complex appearance are single phase α-Ti and they represent the character of allotropic transformations by different routes.

The temperature of 600°C is well below the $\beta/\alpha + \beta$ and $\alpha + \beta/\alpha$ transus for this alloy composition. Thus at the instant of quench, (a) was a metastable β-Ti structure in a α-Ti phase field. It appears that 600°C is also just below the Ms temperature since a group of martensite needles can be seen in the large upper grain. They are characterized by their lenticular shape, spanning a large part of the grain diameter. Because cooling was not resumed, the martensitic transformation halted after this brief burst of effort. This allowed time for competitive nucleation and growth processes to come into play producing the dominant structure of grain boundary envelopes and fine, interleaving lamellar colonies of α-Ti within the confines of the original β grain boundaries. Note that the growth colony dimensions are small in comparison with the original β grains whereas the martensite colonies in (b) are of the same general dimensions as the original β grain diameter.

Etchant: (a) and (b). 20% HF, 20% HNO₃, 60% glycerin. ×200.

CHAPTER V

Diffusion and Transport Processes

Sintering

The term "sintering" is applied to two observable phenomena—the increase in bond strength between two surfaces in contact and the decrease in void content of a porous body or densification. The distinction is significant because the mechanical strength of a powder aggregate can be increased by an order of magnitude with only small change in apparent density. Sintering constitutes a thermal cycle involving time, temperature, and sometimes stress and a gaseous environment. The sintering temperature usually is above that for recrystallization of at least one of the phases present, not because recrystallization itself is part of the process of sintering but because, as in recrystallization, high atomic mobility is a necessary condition. Sintering processes most often relate to strengthening and densification of powder aggregates, but the bonding aspect applies in principle as well to aggregates of short length fibers produced by felting, to clad sheet assemblies produced by pack rolling, and to the increase in cohesive strength of any two surfaces in intimate contact by virtue of thermal treatment in the solid state.

When two surfaces are placed or forced into apparently intimate contact, the real extent of contact is very small. Viewed at high magnifications, any surface will appear undulating or sharply serrated so that mating surfaces can only make real contact at points of asperity. The extent of real contact can be increased appreciably by forcing heavy plastic deformation into both bodies along the plane of contact. In this way point contacts are widened by mutual upsetting into planar interfaces of substantial area. This is the essence of pressure welding, but only with very soft metals can pressure and deformation alone produce appreciable bond strength. In the majority of cases, substantial bonding requires a simultaneous or sequential thermal treatment. Sintering involves the mechanisms by which the voids

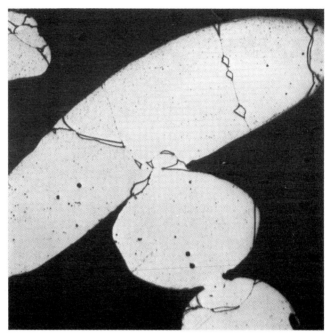

FIG. 5.1. Short-length metal fibers suspended in a fluid medium can be felted as in paper manufacture to produce a porous pad with a multitude of random physical contacts between fibers. If the felt is then heat treated under reducing atmosphere conditions, each physical contact becomes a sintered bond and the felt acquires considerable mechanical strength in spite of the existence of 80–90% porosity. The sintered contacts between fibers of stainless steel are illustrated above.

The extended neck contact between the two circular shaped traces of contacting fibers suggests that the center-to-center distance has actually increased as a result of sintering. This is an illusion brought about by a random section through the extended area of sintered contact which does not truly bisect this region.

It is also interesting to note how grain growth has reached from one fiber across the narrow isthmus of contact to the other. Thus the orientation of austenite in the middle fiber has extended by grain growth into the adjacent one. This is a curious effect because surface tension would tend to hold the grain boundary at the neck to minimize its area.

Etchant: 5 gm $CuCl_3$, 10% HCl, 90% ethyl alcohol. ×250.

existing in the microtopography of mating surfaces are replaced by metal atoms. Evidence exists that this mechanism is vacancy diffusion controlled on the one hand, and a form of surface-tension-activated creep deformation on the other. The truth probably lies somewhere in the middle. That is, under certain circumstances diffusion transport of

FIG. 5.2. Bimetallic and polymetallic clad metals can be made by cold rolling, with a heavy reduction, the two or more sheets that are to compose the composite. The success of this type of bonding depends on cleanliness of mating surfaces, degree of cold deformation in a single pass, and a low temperature anneal for a short time. The above micrograph illustrates a fivefold clad of copper sandwiched between steel in turn coated with aluminum. The bond between these elements is in general as strong as the base metal on either side. Yet there is no visible evidence of interdiffusion.

Etchant: 2% HNO_3, 98% ethyl alcohol, and 33% NH_4OH, 33% H_2O_2, 34% H_2O. ×150.

atoms and lattice vacancies is dominant and in others (notably hot pressing) creep deformation plays the major role.

The actual increase in true area of contact is difficult to appreciate from observations on powder aggregates. The progress of sinter bonding can be more easily appreciated from aggregates of wires or fibers which represent a complex of crossed cylinders. In Fig. 5.2 it can be seen that the original point or line contact has been broadened to finite width. Clearly this has involved considerable mass transport. Yet the bonding aspect of sintering does not need to involve substantial mass transport. Attention is directed to Fig. 5.2 which represents a multilayered clad which has been produced by a cold rolling operation followed by a short cycle heat treatment. In spite of the potential alloying activity of the elements involved and in spite of the apparent perfection of the interface, interdiffusion is not visible.

The progression of consolidation of copper powders under the joint action of heat and pressure can be seen in Fig. 5.3. At low tempera-

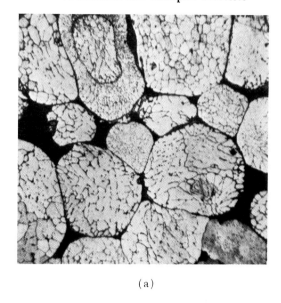

(a)

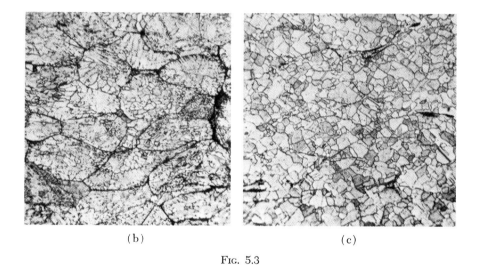

(b)   (c)

FIG. 5.3

tures, the powder aggregate is clearly coherent by virtue of particle-to-particle pressure welds. The individual identity of the powder particles is retained even to the point of vanishing porosity.

The final elimination of all porosity by thermal treatment alone is apt to be a very protracted process. This is perhaps where the action of diffusion processes can be most easily appreciated. The grain boundary plays a vital role in densification. In the later stages of densification it will be observed that the remaining porosity resides in the interior of grains. Almost none will be intersected by a grain boundary. The area left by an advancing grain boundary has a lower void population than regions ahead of the line of advance.° These observations support the proposition that densification proceeds by "evaporation" of vacant sites from pores and that the grain boundary network is a pipeline down which vacancies can rapidly migrate to the external surface of the body. Under conditions of arrested grain growth, voids lying in grain boundaries and in the near vicinity will disappear first while those in the grain interior will persist for very long times. By the same token, if grain growth proceeds too quickly, many voids will be by-passed by migrating grain boundaries leaving a condition which makes full densification prohibitively difficult. The situation can be modified by choosing particle sizes which give a fine pore dstribution which in turn effectively inhibit grain growth. Figure 5.4 illustrates the ability of pores to restrain the migration of grain

° See B. H. Alexander and R. W. Balluffi, *Acta Met.* **5**, 666–677 (1957).

FIG. 5.3. Structures developed by hot pressing atomized copper powder at various temperatures and pressures: (a) 500°F, 5000 psi, (b) 1000°F, 5000 psi, (c) 1800°F, 20000 psi. Atomized copper powder is spherical, and so the mechanical interlock factor in cohesion is largely missing. At low pressures and temperatures, the original shapes of the powder particles is hardly changed except for the upsetting at the original points of contact. In this condition the sintered body is made up of essentially pressure welds. One can easily see the dendritic character of the grains in each powder particle demonstrating the freezing process origin of the powder.

Although the density of (b) is about 95% of theoretical, the individual powder particles have retained their identity. Recrystallization has occurred within the individual particles but only at the highest temperature has it crossed over original particle interfaces. At this state, the only vestige of the powder origin is the vague network of insoluble inclusions which once were oxide films on powder particles.

Etchant: (a)–(c). 33% NH_4OH, 33% H_2O_2, 34% H_2O. (a) ×170, (b) and (c) ×68.

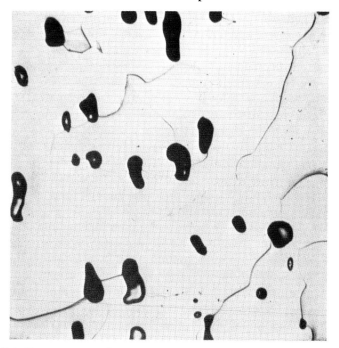

FIG. 5.4. This specimen of porous tungsten has an involved thermal and composition history. Originally it was a dense, two-phase aggregate as illustrated in Fig. 1.11(a) with individual grains of tungsten surrounded by a network of a Ni-Fe-W alloy. It was rolled into sheet as illustrated in Fig. 1.11(b). A zone-sweeping operation was then performed by which a narrow hot zone (above the m.p. of the envelope phase) was walked from one end of the sheet to the other. The majority of the low melting phase was swept out. Only a very small proportion is visible in the resultant structure. As the melt zone swept by each point, existing undissolved particles of tungsten grew as seed nuclei to form the polycrystalline aggregate populated with large pores shown here. The temperature being of the order of 1500°C, grain growth proceeded as soon as the melt zone moved on. The advancing grain boundaries were held back by the pores as can be seen by the curvature in the grain boundaries held between two pores.
Etchant: 5 gm $K_3Fe(CN)_6$, 5 gm NaOH, 100 ml H_2O. ×250.

boundaries. Revived densification can also be attained by introducing cold work (repressing) into the powder aggregate so that a new generation of grain boundaries migrates through the structure.

Intermetallic Diffusion

Diffusion associated with nucleation and growth processes in either the liquid or solid state is not observable by micrographic examination. In general the dimensional scale is too small although the homogenization of a cored structure for example can be followed in a qualitative manner. Metallographic observations, however, can be useful in macroscale interdiffusion processes such as cementation, hot dipping, and sintering of powdered metal mixtures. In such circumstances, the creation and growth of strata of phases intermediate in phase equilibria between the initial components can be observed.

When two pure metals in intimate contact are induced to interdiffuse by heating at elevated temperatures for prolonged periods, a series of parallel or concentric strata are formed which correspond in identity to individual intermediate phases existing in binary phase equilibria at the diffusion annealing temperature. This may be seen in Fig. 5.5. The thicknesses of each layer of intermediate phase are not equal and, in some cases, a layer may be so thin as to escape all but the closest scrutiny. Presumably the thicknesses of these strata lying transverse to the axis of interdiffusion reflect the relative rates of atomic mobility through each of the intermediate phases. On the basis that the rate of transfer of any one atom species must be approximately the same at each interface, only those phases where atomic mobility is high can produce a thick layer and those in which atomic mobility is low must be limited to very thin layers. The formation of a diffusion layer is both a nucleation and growth process as evidenced by Fig. 5.6. Because of this, certain strata may not exist in the early diffusion stages.

Two-phase strata in binary element interdiffusion cannot exist at the temperature of heating primarily because a binary alloy system does not allow the existence of concentration gradients in a two-phase structure. Diffusion can only proceed in the presence of concentration (or more correctly chemical potential) gradients. Two-phase structures can be observed, however, as a result of crystallization if one of the strata is a liquid or of precipitation if one of the solid phases is supersaturated or metastable on cooling.

While interdiffusion proceeds at elevated temperature the parallel or concentric layers of intermediate phases are in close contact and maintain true interfaces. At room temperature the diffusion zone is often observed to be exfoliated or riddled with cracks. This is due to the incompatibility of thermal coefficients of contraction and the nonplastic nature of the phases formed. Coherency of surface cemented coatings is one of the major practical problems in their useful development. Not all types of cracking, however, are unwelcome. In the development of oxidation-resistant coatings a fine grid pattern of cracks

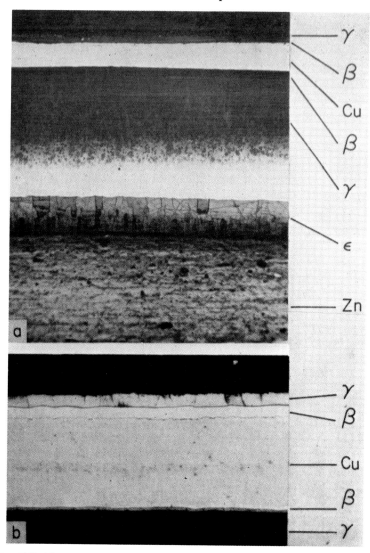

Fɪɢ. 5.5. These companion micrographs derived from a specimen of Zn sheet roll clad on both sides by a thin cladding of Cu. In the initial condition Cu was pressure bonded to the Zn and no interdiffusion had occurred. A portion of the composite sheet was encapsulated in an evacuated glass bulb and heated at 350°C

running transverse to the plane of the coating but not in the plane of the coating may prevent further peeling by subsequent thermal cycling. Provided that the oxidation products fill the cracks and preclude further oxidation, the existence of a particular arrangement of cracks can be advantageous.

Interdiffusion involving the transport of three atomic species can be much more complicated. In principle one would expect parallel or concentric layers of all single and two-phase fields lying in an isothermal ternary section along a line joining the interdiffusing components. Clark and Rhines° have shown that this is not the case. Phase mixture strata can appear which do not lie on a line joining the initial components of the diffusion couple in the isothermal section of a ternary alloy system.

The practicalities of producing useful clad bimetallic materials require an appreciation of the likelihood of the formation of brittle intervening layers. If these occur the clad plate or wire will not be amenable to forming or cutting without internal rupture. To preclude

° J. B. Clark and F. N. Rhines, *Trans. ASM.* **51**, 199–217 (1959).

---

for 1 day. During this period two processes of diffusion transport were operative. Zn and Cu interdiffused across the original pressure bonded interface and Zn vapor diffused into the free surfaces of Cu.

In the region of the original Cu/Zn interface (a) diffusion produced parallel layers representing in sequence the single phases, ϵ, γ and β, which belong to the Cu-Zn alloy system. In addition the portion of Cu near the β interface has the brass color of the Cu-Zn terminal solid solution. The low magnification is necessary to gain the panoramic view of all layers although at this magnification the β layers are not resolvable.

The interdiffusional relationships between Cu and Zn vapor produce only the γ and β layers as seen in (b). This indicates that the vapor pressure (or more properly the fugacity) of Zn over the ϵ phase is higher than over pure Zn. As a result the ϵ phase is not formed. True equilibrium is established between the γ phase and Zn vapor (see F. N. Rhines, "Surface Treatment of Metals," p. 131. ASM, Metals Park, Novelty, Ohio, 1941). An alternate explanation is that ϵ cannot easily be nucleated by Zn vapor on a γ brass surface.

The low magnification micrograph illustrates the difficulty in clear etching of a polyphase structure in an electrolyte where large electrochemical potentials exist between phases. For example, the strongly anodic behavior of γ in electrical contact with β causes gross over etching of the former in the time necessary to resolve the structure of the other layers.

Etchant: (a) and (b). 33% NH$_4$OH, 33% H$_2$O$_2$, 34% H$_2$O. (a) $\times$100, (b) $\times$500.

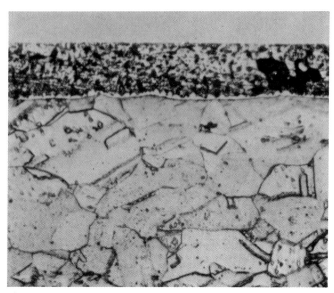

FIG. 5.6. This is a section from a Cu component which had been hot dipped in a Pb-Sn solder bath. Immersion in the molten solder was long enough to allow diffusion of Sn into Cu forming a layer of intermetallic compound. This specimen shows the early stages of layer growth. The scalloped contour on the solder side of the couple indicates that the intermetallic compound nucleated at separate points and grew radially in all permissible directions. The lateral directions of growth gradually impinged cutting off all remaining contact between the molten solder and solid Cu. At this time the solder was observed to de-wet which was the nature of the practical problem at hand. This case serves to emphasize that change in apparent wetting is usually associated with the intervention of a third component in a two-component system. Sometimes the new component is a re-formed oxide but just as commonly it can be an intermetallic compound layer.

Etchant: 33% NH_4OH, 33% H_2O_2, 34% H_2O. $\times 1000$.

FIG. 5.7. Cladding steel directly with Ti gives an unsatisfactory product because the interdiffusion of Ti and Fe yields a series of layers of brittle intermetallic compounds. The key to the successful development of a clad which can tolerate metal working operations is the choice of intervening thin layers of noninteracting metals. For example, V and Ti interdiffuse slowly but the range of solid solutions is ductile so that the two metals bonded together are always ductile. Likewise at elevated temperatures V and Fe form a continuous series of solid solutions. However, on cooling through lower temperature ranges a sigma intermediate phase can form. Also V can take carbon from the steel to form stable carbides. A trimetal clad, Ti-V-steel, using V as an interliner is shown in (a). The sigma phase plane and some carbides which form at the V-steel interface are almost continuous but their detriment to the ductility of the clad is not very serious. Complete freedom from brittle phases can be achieved by the use of a four-ply construction Ti-V-Cu-steel (b). Each of these metals is compatable with its immediate neighbors.

Etchant: (a) and (b). 20% HF, 20% HNO_3, 60% glycerin. (a) $\times 225$, (b) $\times 90$.

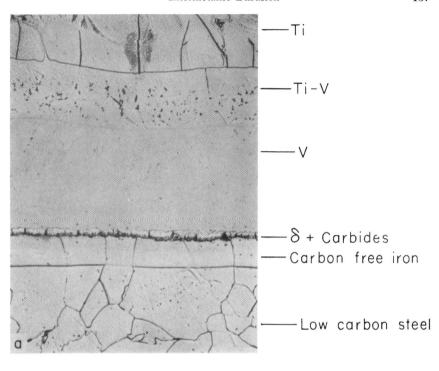

— Ti

— Ti - V

— V

— δ + Carbides
— Carbon free iron

— Low carbon steel

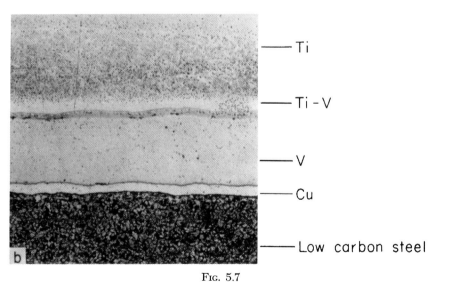

— Ti

— Ti - V

— V

— Cu

— Low carbon steel

FIG. 5.7

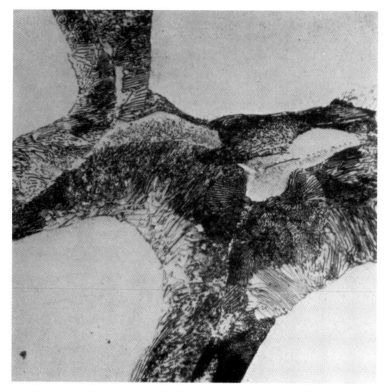

FIG. 5.8. Molten Cu in a brazed joint between two pieces of low carbon steel produces a carbon segregation in the steel in the following manner: Cu dissolves into the surface of the steel (austenitic at the brazing temperature). The Cu dissolved in the austenite radically changes the chemical potential of carbon in austenite leading to carbon segregation to the Cu-rich surface. Copper itself will rapidly penetrate carbon-rich austenite intergranularly. This artificial carbon surface-enrichment leads to intergranular penetration by Cu which in turn, as it dissolves into the grains on either side of each grain boundary, causes further carbon segregation. The above micrograph verifies the steep uphill diffusion of carbon to the Cu-rich grain boundaries because the transformation structure is almost all pearlite as compared to the carbide-free grain centers.

Etchant: 2% HNO_3, 98% ethyl alcohol. $\times 1000$.

this, one can design composite systems with thin diffusion barriers which permit compatibility for the lifetime of the component to be made. An example of this is illustrated in Fig. 5.7.

One is accustomed to think of diffusion from points of high con-

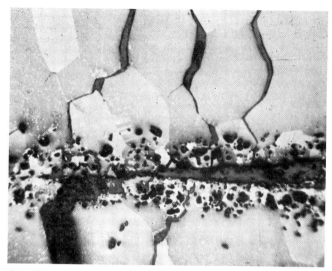

Fig. 5.9. This specimen was originally a composite made by roll cladding of Zn between two sheets of Cu. The as-rolled triclad was annealed at 400°C for 7 days. The Zn atoms being more mobile than the Cu atoms, large numbers of vacant sites were formed in the Zn which ultimately condensed into void pockets. As the γ brass interface moved in, it passed through the porous field incorporating the individual pores into the centers of large grains. In the region of densest population of voids, these act as grain growth inhibitors. The generation and disappearance of various phases must cause local stresses to exist, for there is considerable intergranular cracking in the final γ brass structure.

Etchant: 2 gm FeCl₃, 10% HCl, 90% H₂O. ×50.

centration of a given atomic species to points of lower concentration. Occasionally this generality is violated and the by-product of diffusion is the segregation of some element which was previously uniformly distributed. A case in point is the segregation of carbon in the region of the brazed joint illustrated in Fig. 5.8. This example serves to emphasize that diffusion occurs in the direction of a gradient of chemical potential even though this gives the appearance of uphill diffusion. In the instance illustrated, the diffusion of copper into the low carbon austenite, lowered the chemical potential of carbon so that carbon from the copper-free austenite diffuses into the copper-rich austenite regions to equalize the chemical potential. In so doing the carbon content of the copper-rich regions increased to approximately eutectoid proportion.

When the mobility of one of two interdiffusing atomic species is appreciably greater than the other, the differential mass transfer must be compensated by the formation of vacant sites and their drift in the direction opposite to that of the faster moving species. Vacancies can be considered a sort of alloy component in that supersaturation can occur and must be relieved. The precipitation of vacant sites is a nucleation and growth process which leads to the formation of visible pores. A case of diffusion-produced porosity is illustrated in Fig. 5.9. In this instance, another metal phase is the sink for the more highly mobile zinc atoms. Volatilization to the surrounding atmosphere as with a brass heated in vacuum can produce the same effect when the rate of departure of zinc atoms from the free surface is greater than the rate of supply by solid state diffusion.

Liquid Metal Penetration

The surface tension relationships between a liquid and a solid and between liquid, solid, and gas can lead to a variety of events, some

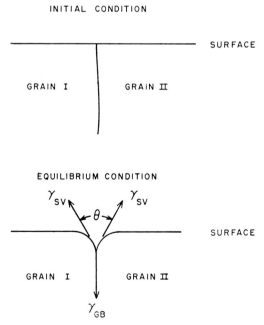

FIG. 5.10. Diagrammatic illustration of the origin of grain boundary grooving at a free surface as a result of the equilibrium between the surface tension forces of the solid in its own vapor, γ_{SV}, and the grain boundary tension, γ_{GB}.

useful and some destructive. In this context must be included liquid phase sintering, infiltration of porous bodies, brazing, and intergranular penetration. Consider an ideal planar surface to which emerges a grain boundary at an approximately perpendicular angle. As at all faces and interfaces, a surface tension exists proportionate to the specific energy of each of these. At the point of emergence, three tension vectors exist: γ_{GB} and two γ_{SV} where the subscripts GB and SV denote the grain boundary interface and the solid-vapor face. The initial geometry is thermodynamically metastable with respect to the groove illustrated in Fig. 5.10. The equilibrium groove angle, θ, is defined by simple triangle of forces:

$$\gamma_{GB} = 2\gamma_{SV} \cdot \cos \theta/2$$

The grain boundaries of iron revealed in the hot stage microscopy view of Fig. 5.11 are the result of thermal grooving brought about by surface tensional forces.

If the solid-vapor face is replaced by a solid-liquid interface, a

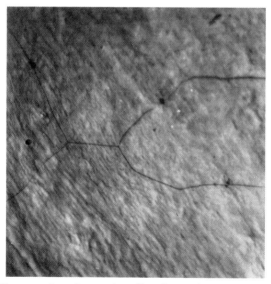

FIG. 5.11. The grain boundaries of unalloyed iron shown here are the result of thermal grooving. The specimen in the polished and unetched condition was set in the heated platform of a hot stage microscope. The specimen experienced a thermal excursion to 780°C. This micrograph was photographed while the specimen was at this temperature.
Thermally etched. ×350.

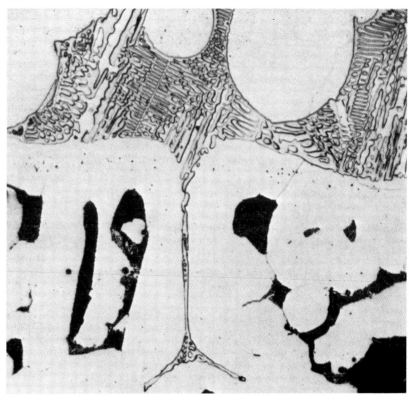

Fig. 5.12. This is the interface of a joint of SAE 1045 steel brazed with a eutectic Fe-B alloy. The eutectic melt wets the steel readily and, having a low dihedral angle as may be seen by the surface groove, it travels rapidly up the boundaries of the austenite grains existent at the brazing temperature. Given sufficient time the liquid would have enveloped the surface grains. The micrograph shows considerable progress in this direction already. The eutectic melt itself is unsaturated with respect to iron at the brazing temperature and proceeds to dissolve some from the base metal. This iron as may be seen is crystallized out as large primary dendrites during freezing of the joint.
Etchant: 15% HNO₃, 15% HCl, 70% H₂O. ×1000.

similar change in topography must occur as governed by the analogous triangle of forces:

$$\gamma_{\rm GB} = 2\gamma_{\rm SL} \cdot \cos \theta/2$$

As demonstrated in Fig. 5.12, liquid metals can do more than simply establish a groove at every emerging grain boundary. They can lit-

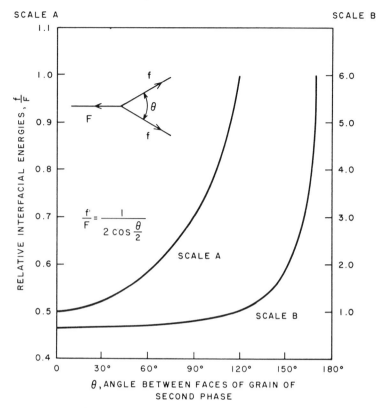

FIG. 5.13. Correlation between interfacial energies of two phases and their equilibrium dihedral angle.

erally travel up grain boundaries and isolate whole grains from the remaining aggregate. The conditions for this are simply defined. Penetration along a grain boundary by a liquid involves the loss of unit area of grain-to-grain interface and the gain of two unit areas of solid-liquid interface. When the change of interfacial energy $\Delta\gamma = \gamma_{GB} - 2\gamma_{SL}$ is positive, the intergranular penetration by the liquid is favored. The critical condition, therefore, is: $\gamma_{SL}/\gamma_{GB} \leq \frac{1}{2}$. This also indicates the magnitudes of the dihedral angle θ, associated with the grain boundary penetration. The graphical relationship between γ_{SL}/γ_{GB} and θ shown in Fig. 5.13 is taken from Smith[*]; it can be seen that $\gamma_{SL}/\gamma_{GB} = \frac{1}{2}$ is approximately satisfied for values of θ less than 60°.

[*] C. S. Smith, *Trans. AIME* **175**, 15–51 (1948).

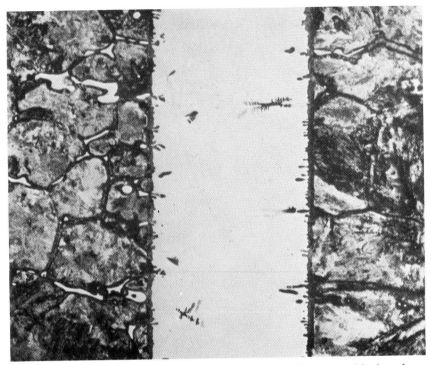

FIG. 5.14(a). This micrograph covers the region of a butt joint of high carbon steel (0.9% C) brazed with pure Cu. The total time of the brazing operation is only a few minutes, yet in that time appreciable grain boundary penetration (0.018 in.) has occurred. Moreover, the intergranular (austenite) penetration by molten Cu was followed by diffusion into the austenite grains leaving dark etching bands at the prior austenite grain boundaries and isolating islands of remaining Cu. The Cu of the brazed joint has also dissolved appreciable Fe which almost quantitatively crystallized out when the joint cooled. The crystallization of Cu-saturated Fe occurred both as well developed primary dendrites and as oriented growths at the Cu/steel interface. The anisotropy of growth rates from a melt is clearly demonstrated by the dendrite arms growing ahead of the thin band of Fe which initially formed at the interfaces. Notice also that the dendrite arms are all parallel indicating that the austenite grains of the solid surface are not nucleation sources for austenite crystallizing from the melt.

Etchant: 2% HNO$_3$, 98% ethyl alcohol. ×135.

Spreading by a liquid over a solid metal surface is also governed by the inequality of associated surface energies:

$$\Delta\gamma = \gamma_{SV} - (\gamma_{LV} + \gamma_{SL})$$

where LV denotes liquid-vapor.

Spreading proceeds when the conversion of unit area of free surface

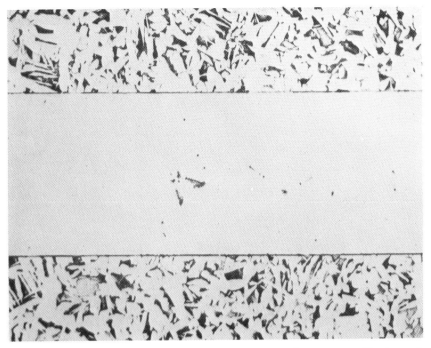

Fig. 5.14(b). This is a micrograph of a butt joint between two like bars of low carbon (∼0.2% C) steel brazed with pure Cu. This micrograph is included for comparison with Fig. 5.14(a) to demonstrate the influence of carbon content on the grain boundary penetration capability of molten Cu. There is no apparent penetration of Cu around prior austenite grain boundaries.
Etchant: 2% HNO_3, 98% ethyl alcohol. ×135.

to approximately unit area each of liquid-solid interface and liquid-vapor interface yields a finite loss of internal energy to the system, i.e., $\Delta\gamma > 0$. Spreading is complicated by the presence of oxide films which in normal brazing practice must be removed by the action of molten salt fluxes or reducing atmospheres.

Minor elements both in the solid and in the liquid can be a major factor in the magnitude of the dihedral or wetting angle. Both increasing carbon content in the steel [Fig. 5.14(a,b)] and minor amounts of lithium in the melt strongly influence the wetting of steel by molten copper. Wetting is often accompanied by dissolution of the base metal until equilibrium is established as shown in Fig. 5.15.

Intergranular penetration can be more rapid than composition equilibration as shown in Fig. 5.16 where ultimate dissolution of the

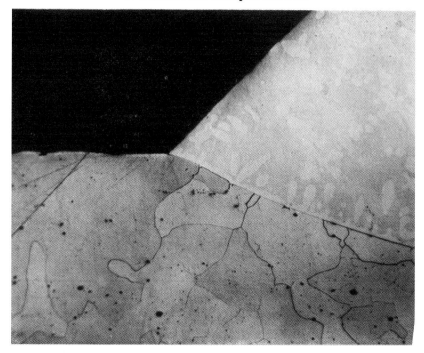

Fɪɢ. 5.15. When the wetting angle is large and intergranular penetration is negligible, and the melt is unsaturated with respect to the base metal, simple dissolution will proceed carving into the base metal. The depth of penetration is roughly proportional to the depth of melt immediately above. These points are illustrated by the section of a sessile drop of a Ni-Si alloy resting on a plate of Armco iron.

Etchant: 2% HNO₃, 98% ethyl alcohol. ×135.

base metal by the penetrating liquid resulted in a change of dihedral angle sufficient to cause de-wetting along the grain boundaries.

Complex structures composed of high and lower melting phases can be synthesized by two different powder metallurgy approaches. By one method, a porous, sintered skeleton of the higher melting phase is infiltrated by liquid metal. This involves capillary transport of liquid through interconnecting voids. Infiltration permits densification without large dimensional changes from the original skeletal shape. The dimensional control depends on the retention of original sintered bonds. This is only permissible if the dihedral angle between the solid and liquid is greater than 60°. Thus for infiltration purposes too effective wettability is not desirable. The prerequisite of dimen-

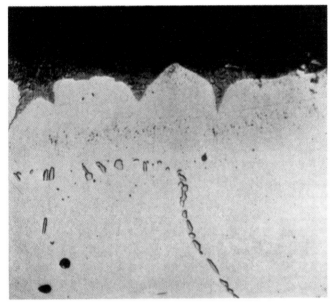

FIG. 5.16. It can happen that when a liquid penetrates the grain boundaries, it changes composition by absorption of the surrounding metal. In so doing, the melting point is raised and so the liquid film freezes although the temperature has not changed. It may also happen as illustrated here that absorption increases the interfacial energy causing the liquid to de-wet. The film now becomes discontinuous although still disposed along grain boundaries. Melt phase: Ni-B eutectic. Base metal: type 304 austenitic stainless steel.
Etchant: 15% HNO_3, 15% HCl, 70% H_2O. ×750.

sional control carries with it the price of incomplete densification because, with the limited wettability, it is not possible to gain access to unconnected pores. It can be seen that the influence of carbon in steel on the intergranular penetration into austenite, which is an important consideration in brazing, is for the same basic reasons important in copper infiltration of iron powder sintered bodies.

In other applications of powder metallurgy, the need is not so much for controlled shape as for complete densification and for an ideal microstructure in which one phase in fine particulate form is enveloped by another. The purpose of the envelope structure is usually to provide toughness or special electrical properties to very hard or very high melting point substances. Such is the case for cobalt-bonded tungsten carbide and silver-bonded tungsten. The envelope character of the ultimate structure demands that the dihedral angle between

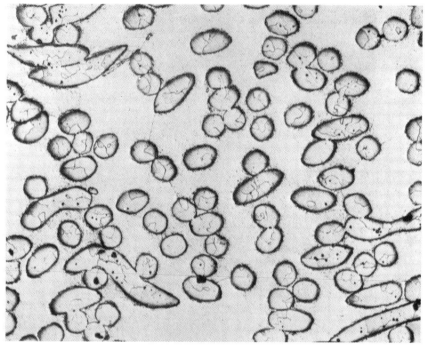

Fig. 5.17. The structure shown was produced by sintering a mixture of 33% Fe in short length fiber form and 67% Ag above the melting point of the latter. The solubility of Fe in molten Ag is negligible so the fibers retain their original shape but the dihedral angle between molten Ag and solid Fe is large enough to induce sinter bonding of Fe particles to each other. It will be noted that wherever two particles have bonded over a resolvable area, grain growth has swept across the bonded region so that a grain originally in one particle has grown into the other. Thus the migration of a grain boundary across an interface as a vehicle for elimination of microvoids at an interface serves both in so-called liquid phase sintering as well as the more common condition where the voids between particles are filled with an inert gas or evacuated.

Etchant: 2% HNO$_3$, 98% ethyl alcohol plus 5 gm CrO$_3$, 100 ml H$_2$O, used electrolytically. ×90.

liquid and solid be less than 60°. The high wettability of the liquid makes it unnecessary to have a presintered skeleton of the high melting phase to achieve high density. In fact, the action of the liquid phase in these circumstances would be to destroy the bond areas of the skeleton. Complete densification can be achieved by heating the pressed powder mixture to above the melting point of one constituent. This is commonly called liquid phase sintering.

In liquid phase sintering, voids are eliminated rapidly with large

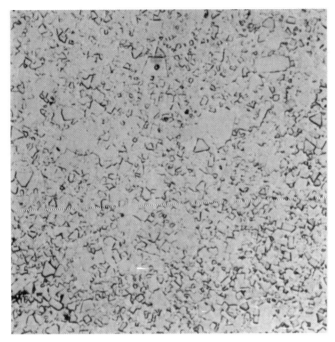

Fig. 5.18. A common grade of sintered carbide for cutting tool applications contains about 6% Co by weight and 94% of a W-Ta carbide mixture produced by the carbon reduction of mixed oxides of W and Ta. During the sintering operation, the Co melts and dissolves an appreciable amount of the carbides. The melt volume is considerably increased and as a result porosity is completely eliminated. The undissolved carbides are approximately spherical nuclei for crystallization of carbides on cooling from the sintering temperature. The final shape of the carbides in the Co matrix is governed by conditions of crystallization from the melt. The angular carbides in the micrograph above illustrates the anisotropy of crystallization growth rates. More intricate dendritic forms are probably precluded by growth interference by the high density of growth centers (undissolved carbide particles). The morphology of carbides in this manner can be influenced by surface energy conditions as may be brought about by interface absorption of minor components. For example, the existence of small amounts of oxygen in TiC-Ni matrix structures can change the carbide form from sharp and angular to rounded and nodular which indicates that oxygen reduces the anisotropy of interfacial energy.

Etchant: 5 gm $K_3Fe(CN)_6$, 5 gm NaOH, 100 ml H_2O. ×1000.

consequent shrinkage. The void elimination can be either the result of surface tension forces or the rapid transport of vacancies through a continuous and sustained path of liquid. Liquid-solid interfacial tensions govern whether the particles of solid agglomerate or disperse. In Fig. 5.17 the dihedral angle between the melt and solid phases is somewhat greater than 60° with the result that the particles of the solid phase are beginning to bond. The structure produced by liquid phase sintering depends considerably on the amount of the solid phase taken into liquid solution during the sintering operation and subsequently rejected on cooling. With the freezing on of new solid, strong anisotropy of crystallization can produce sharply angular particles of the dispersed phase as shown in Fig. 5.18. The solubility of the solid in the melt may be an important factor in the particle size of the dispersed phase.

Gas-Metal Reactions

Examination of metals exposed to gases at elevated temperatures reveal one or more resulting changes of state: (a) The formation of one or more new solid species disposed as layers between the metal and the gaseous environment. While oxidation or scaling is probably the most commonly encountered example of this, the generation of sulfide (by reaction with H_2S), carbide (by reaction with CH_4), or silicide films (by reaction with $SiCl_4$) are equally representative of this change of state. (b) The absorption into solid solution of a component of one of the gas molecular species in the atmosphere. The absorption and inward diffusion sets up a concentration gradient from saturation at the surface to the initial condition somewhere in the interior. Carburization of steel is the most commercially important example of this process. Oxygen and nitrogen at low partial pressures are absorbed in this fashion by the reactive transition metals of Group IV and V. (c) The combination of surface layer formation and subscale absorption. (d) Simultaneous absorption and precipitation of a new phase containing the migrant element. Hydride formation in the Group IV metals falls in this category. So also does internal oxidation which is a special phenomenon in certain dilute solid solutions. The commercial process of nitriding of steel can be regarded as equivalent to internal oxidation. (e) Selective evaporation or reaction with the environment to produce a volatile phase whose removal depletes the metal superficially. Annealing of brass in a neutral or reducing atmosphere can lead to superficial loss of zinc with consequent change in color from yellow to red. Some of the Group V metals, notably niobium and tantalum, can be purified of oxygen simply by heating to very high temperatures in dynamic vacuum. The oxygen literally evaporates from the metal. As in decarburization of steel, Fig. 5.19, the removal of an alloying element may require the cooperation of the gas in the surrounding atmosphere. The removal of carbon from the surface of steel requires interaction with hydrogen or with CO_2. A variation of this is the blistering of copper during annealing in a hydrogen-bearing atmosphere. Under such conditions, hydrogen diffuses into the copper and, by reducing oxide inclusions, generates water vapor inclusions as replacement. These exert great pressure on the surrounding metal, leading to yielding and, near the surface, to the appearance of blisters.

The multilayer formation in oxide scales is rarely as cleanly arranged and clearly resolvable as with intermetallic diffusion. Suboxides may be continuous with the metal surface but the outer oxide is rarely so. This is because of the great differences in specific volume between a metal and its oxides. Oxide growth nucleates at many points simultaneously on the metal surface. These grow laterally in the plane of the interface and also projecting out from the interface. Stresses developed between growth centers lead to frequent cracking of the scale

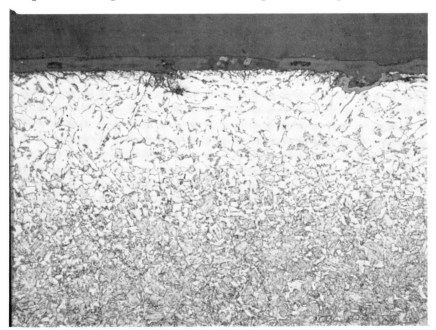

FIG. 5.19. This is an example of surface decarburization of a steel (AISI 8642), which occurred during austenitizing using an improperly regulated "protective" atmosphere. The microstructure was examined after tempering. The surface is largely ferritic with small zones of tempered martensite. The interior is homogeneous tempered martensite as it should be. This product will show inferior wear resistance and fatigue life.

Etchant: nitol. $\times 400$.

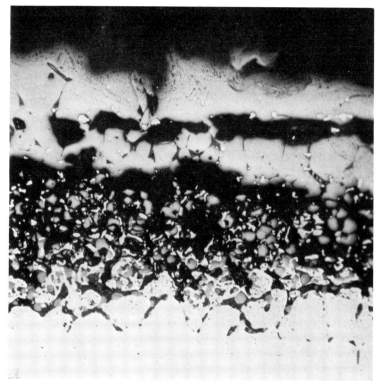

Fig. 5.20. Generally it is very difficult to prepare a cross section of an oxidized specimen preserving the oxide scale and the oxide/metal interface. There are invariably large volumetric changes associated with the conversion of metal to oxide and this, combined with the nonplastic character of oxides, makes for loosely adherent oxide overgrowths. Adherency decreases with increasing film thickness. Because of the repetitive sequence of growth, expansion, and fracturing of oxide films, the simple layer-upon-layer pattern of intermediate phases common to interdiffusion systems is rarely encountered in oxidation growths or scale. Furthermore, the advancing front of oxidation is not usually planar.

These effects are illustrated in this micrograph representing a bar of low alloy steel which has been subjected to air oxidation at about 2200°F for a prolonged period of time. Note that two oxide phases are resolvable (probably Fe_3O_4 and Fe_2O_3); that oxidation penetrates the metal as finger growths enveloping fragments of metal which are probably grains of austenite; that small pieces of unoxidized metal persist even in the separated regions of the scale; and that the two oxides and metal exists near the metal surface as a jumbled mixture.

Unetched. ×150.

and some sort of re-initiation of oxidation at the roots of these cracks. Thus, as shown in Fig. 5.20, oxides scales do not possess very symmetric structures.

As remarked earlier, internal oxidation is peculiar to certain dilute solid solutions. In these cases the minority solute element is prone to form much more strongly bonded oxides than the solvent element.

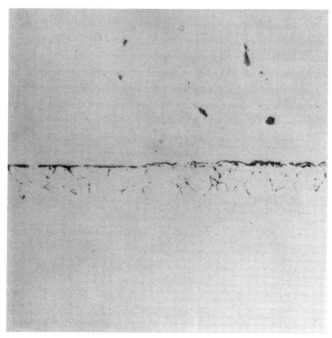

FIG. 5.21. The lower portion of this micrograph represents the surface of a carburized steel specimen preserved from spalling and rounding by a heavy layer of electroplated Ni (upper portion). The steel is SAE 8620 which has been carburized in an endothermic gas atmosphere which contains an appropriate ratio of CO to CO_2 and some nitrogen, and water vapor. Only in the unetched condition can one see the network of inclusions, presumably oxides, which form an intergranular envelope about the austenite boundaries at the time of carburizing. Distributed in this manner they are very injurious to fatigue strength. This represents a special form of internal oxidation characteristic of dilute solid solutions wherein one or more solute elements have the ability to form much more strongly bonded oxides than the solvent metal. Cr and possibly Mn serve the purpose in this instance. This type of intergranular oxidation can be avoided by carburizing in hydrocarbon gases diluted with a neutral gas carrier and free of oxygen. Nominal composition of SAE 8620 steel: 0.18–0.23% C, 0.70–0.90% Mn, 0.40–0.60% Cr, 0.40–0.70% Ni, 0.15–0.25% Mo.
Unetched. ×500.

The absorption of oxygen by the solvent leads to what is literally solid state deoxidation by the minority solute element. The products of deoxidation are distributed as finely divided particles often (as in Fig. 5.21) selectively disposed at grain boundaries. When the concentration of the solute element is raised, the probability of an oxygen atom penetrating the surface to a depth of more than a few atomic distances without encountering an atom of the solute element is so small that internal oxidation does not occur. Instead the solute element forms an almost pure oxide scale at the metal surface.

The formation of coatings by freezing on, by condensation, and by electrodeposition can by careful observation be shown to nucleate at random points on the metal surface and to grow outwards as well as laterally until surface coverage is complete. This seems to happen also in the formation of an oxide film as demonstrated in Fig. 5.22.

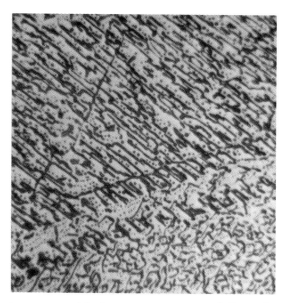

Fig. 5.22 A specimen of high purity, coarse grained Fe in the polished and etched condition was heated in a hot stage microscope through the α to γ inversion temperature and then cooled. During this time it was under a supposedly protective vacuum. There was sufficient air leakage to permit the initiation of oxidation at the surface and from the arrangement of the oxide particles which have incompletely covered the metal surface, it can be seen that the oxide film grows from a number of nucleation sites and that the growth pattern itself is probably governed by the crystallography of the substrate metal grains.

Thermally etched. ×350.

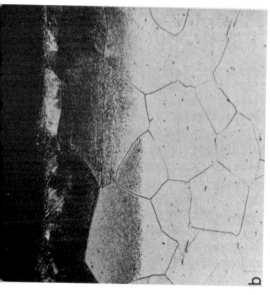

Fig. 5.23. Titanium alloys at elevated temperatures and exposed to air form a rather loose white or yellowish white scale which is primarily TiO₂. This itself is not as pernicious as the penetration of oxygen into the metal itself immediately under the scale. The absorption of oxygen by the surface metal manifests itself metallographically in the two ways illustrated. In either case the result is severe embrittlement.

Structure (a) is of an alloy containing 15% Cb and 10% Al which has been exposed to air at 1000°C for 100 hours. The structure at that temperature was the bcc β phase. The absorption of oxygen raised the β/α + β transus so drastically that the α phase precipitated out in the coarse plates observed. The introduction of oxygen literally changed the constitution of the alloy at the surface.

Structure (b) is of a Ti alloy containing 13% V, 11% Cr, 3% Al. In this case the contaminated surface is residual from hot rolling. The amount of oxygen dissolved is not so great as to change the phase equilibrium conditions but it is enough to increase the rate of precipitation of α such that while the core retains the β phase on cooling, the surface is constrained to precipitate α-Ti in a very finely divided form. The structure itself derives from the ultimate cooling cycle. At the temperature of hot rolling even with the surface enriched with oxygen, the structure was all β.

Etchant: (a) and (b) 20% HF, 20% HNO₃, 60% glycerin. ×100.

Figure 5.23 provides two examples of means by which the existence of subscale absorption of gaseous components can be detected. In one instance the constitution of the alloy is clearly changed, and in the other, the kinetics of transformation.

Corrosion

The reaction of a metal with its environment leading to the conversion of the metal or part of the metal to inorganic salts is the most general connotation of corrosion. In practice, the definition is usually limited to reactions occurring near room temperature as distinct from oxidation or scaling reactions at elevated temperature. Water is nearly always an essential ingredient of the corrosion environment. Common cases include saline solutions, moist, warm air, and condensate. When the rate of corrosion is very rapid as in strongly acid or basic solutions, the term dissolution may be used instead. In this case the reaction products usually dissolve into the solvent of the corrosive solution. Attack by molten salts probably falls into this category.

Most often the products of corrosion are hydrated oxides which are porous and loose encrustations on the surface of the metal. Their very nature assists the perpetuation of the corrosion process because they serve as natural sites for the absorption of condensed moisture and the solution therein of chloride salts, and various sulfur, carbonic, and ammoniacal gases.

One can encounter various peculiar forms of corrosion of a selective nature, but corrosion is most commonly observed as a general attack. That is, in viewing the cross section of a metal near and at the surface after undergoing corrosive attack, the surface, while no longer smooth, reveals no systematic preferred directions or sites of corrosion. Even though no one specially selected area seems more prone than others to corrosive attack, corrosion processes are generally regarded as electrochemical in nature. By this is meant that electrochemical potential differences between resolvable or irresolvable microstructural features and current flow through the corrosive medium lead to anodic dissolution of the metal. Strength to this argument is gained by the ability to protect a metal in a corrosion media by electrical coupling with a more anodic metal. Thus, for example, zinc and magnesium in electrical contact with iron and immersed in the same corrosive environment prevent corrosion of the latter. Of course, the corrosion rate of the zinc or magnesium is accelerated. This is termed sacrificial corrosion, and it is said that zinc or magnesium provide cathodic protection by rendering the local anodes in the iron cathodic to the sacrificial metal.

Many noble metals are resistant to corrosion in certain environments because the thermodynamic driving force for conversion to an inorganic salt is insufficient. Other less noble metals appear also to be

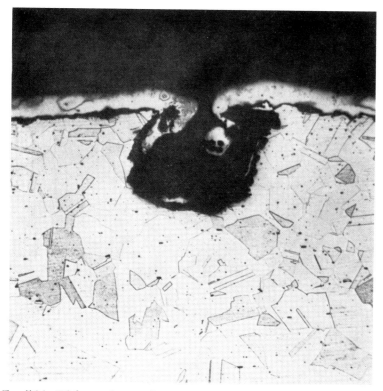

FIG. 5.24. While stainless steel is resistant to general corrosion it has a proneness to a very localized form of corrosion in strong halide solutions. This special type of deterioration is called "pitting corrosion" and it generally occurs in corners, crevices, and at the edges of masked-off areas. The effect can be reproduced in the laboratory by immersing a specimen of austenitic stainless steel which has been only partially coated with an adherent wax in a solution of about 10% FeCl₃. Pitting corrosion will develop just under and along the edge of the waxed-off area. The pit shape bears no relationship to the grain structure as shown in the accompanying micrograph. (The surface coating is of electroplated Ni to preserve the specimen edge during metallographic preparation.) The pit propagates relatively little in width but at a much greater rate in depth indicating that the stagnant conditions of a crevice are conducive to this type of corrosion.

Etchant: 5 gm CrO₃, 100 ml H₂O, used electrolytically. ×150.

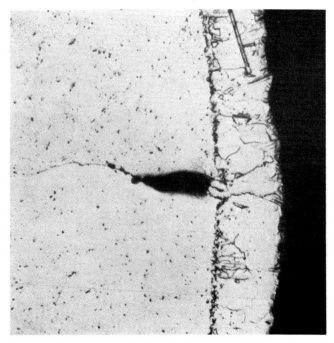

Fig. 5.25. This is a specimen of a very dilute solution of 0.05% As in Pb. The dendritic structure indicates that it is a cast alloy. The specimen has developed intergranular corrosion in simulated lead storage battery service which involved immersion in strong H_2SO_4 solution and an imposed potential of 2.5 volts above pure lead. The intergranular character of the corrosion indicates that in this very dilute solution, As probably segregates to grain boundaries producing a grain boundary region which is significantly anodic compared with the grain interiors.

In order to preserve the surface irregularities during metallographic preparation it is often necessary to electroplate a layer of metal over the original surface. In this case a layer of Pb was electroplated over the corroded surface of the Pb alloy specimen. The choice of Pb as a plating metal minimizes electrochemical processes during etching which mask the structure of the specimen. It is a good requirement of the plating metal that it not interfere with etching. Strongly anodic relationships between the specimen and the coating will produce interference in the form of local exaggerated rates of etching, staining, loss of definition of certain metallographic features.

Etchant: 15% HNO_3, 15% acetic acid, 70% glycerin. ×250.

corrosion resistant but because of the natural existence of a thin, uniform, adherent, and inherently corrosion resistant oxide film. This is the case for stainless steel. The usual moist air environment which causes rapid rusting in ordinary iron and steel is generally inert to stainless steel. This condition can be changed radically if relatively

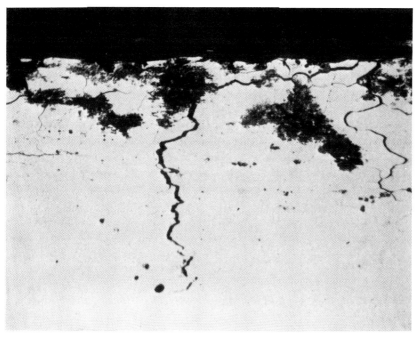

FIG. 5.26. The Al 2024 alloy in its aged hardened state is highly susceptible to both cracking under stress or intergranular corrosion penetration with or without stress. This micrograph illustrates the existence of both active processes as produced in a tensile specimen stressed to 90% of its engineering yield point while immersed in a NaCl–Na$_2$CrO$_4$–HCl solution. Major cracks seem to follow substantially intercrystalline paths roughly transverse to the axis of the principal tensile stress. It is not uncommon, however, to see satellite cracks running in directions parallel to the axis of principal tension. This is usually symptomatic of a highly anisotropic material and the nonuniform distribution of microstresses associated. The black patches are in fact a myriad of fine intergranular separations muddied by corrosion products. Many small grains are literally isolated from the parent body of metal in these regions. This form of intercrystalline penetration proceeds even without the existence of external or internal residual stresses. Nominal composition of Al 2024 alloy: 4.5% Cu, 1.5% Mg, 0.6% Mn.

Unetched. ×250.

small amounts of chloride ions exist in the air-saturated, condensed moisture. The chloride ion is capable of destroying the impermeable character of the chromium-rich oxides which naturally form on stainless steels.

One of the more pernicious forms of corrosion is called "pitting corrosion" because of its highly localized nature (Fig. 5.24). It follows a very localized path in the metal which has no obvious relationships with microstructural features. It is thought that certain ion species can concentrate in physical crevices and induce more rapid corrosion there than elsewhere. Pitting corrosion is therefore self-perpetuating. The reasons for the initiation of pitting corrosion are not always very obvious, but the process can be synthesized in the laboratory by artifices which in fact create the physical characteristics of a crevice.

Any bonded assembly of dissimilar metals is likely to create selective corrosion in the more anodic metal. This is a common dilemma in brazed assemblies where the most suitable brazing material, from the viewpoint of achieving high joint efficiencies, may be highly anodic to the other components of the joint. Even mechanical coupling such as in rivetted joints can produce severe local corrosion of one or other elements of the joint.

Minor alloying elements which cause concentration gradients of certain major alloy species or themselves segregate to grain boundaries can lead to selective intergranular corrosion (Fig. 5.25) if large electrochemical potentials exist between the extremities of the composition gradients. Most often the existence of these composition gradients cannot be recognized a priori by microstructural examination. But putting the case in reverse, the recognition of intergranular corrosion by metallographic means justifies intensive study of the existence of minor impurities and then possible interaction with other alloying elements during thermal treatments. The classic case of intergranular corrosion is brought on in stainless steel by the presence of minor amounts of

---

FIG. 5.27. Brass pipe used as hot water condulet in regions where the mineral content of the water is high suffers a form of failure called "de-zincification." What happens is that the brass dissolves slowly into the water but being considerably less noble than Cu, the Cu ions are displaced from the water by additional dissolution of the brass. The displaced Cu deposits on the brass as a loose metallic deposit. A vigorous cell action is set up between the Cu and the brass which accelerates the corrosion of the latter and the build up of the former. This process can be simulated in the laboratory by immersing a piece of brass in CuCl$_2$ solution acidified slightly with HCl. Over a period of a few weeks, about one half of the thickness of the 1/8 in. brass plate was corroded and replaced by a spongy copper deposit. The external appearance of the specimen was remarkably unchanged except for the Cu color on the surface.

Two views of this specimen are shown here; one in the etched condition (a) to reveal the striated structure of the Cu deposit which looks very much like an electroplated material; and one in the unetched condition (b) to demonstrate how etching which reveals grain structures also exaggerates the size of voids and pores to a grossly distorted state.

(a) Etchant: 33% NH$_4$OH, 33% H$_2$O$_2$, 34% H$_2$O. ×50.

(b) Unetched. ×50.

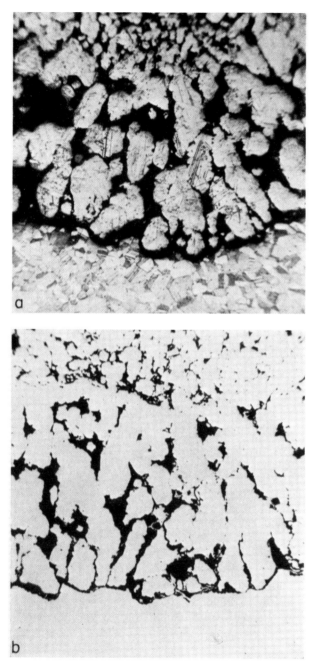

Fig. 5.27

carbon. During thermal cycling the intergranular precipitation of chromium carbides leads to local depletion of chromium in the immediately adjacent austenite. The electrochemical differences between Cr-rich and Cr-poor austenite is sufficient to general severe intergranular corrosion in aerated saline solutions.

A mysterious form of failure in metals occurs sometimes in the presence of certain aqueous or gaseous environments and with the simultaneous existence of stress either externally applied or residual from prior thermal-mechanical history. This type of failure is termed "stress-corrosion cracking" (see Chapter II). In stainless steels this type of cracking looks like brittle fracture with no apparent corrosion. In aluminum alloys it is almost always associated with intergranular corrosion, as in Fig. 5.26

In brass a peculiar form of reversible corrosion can occur. Here the corrosion products which are soluble salts of copper and zinc react with the remaining brass to reprecipitate the copper in a spongy form (see Fig. 5.27).

CHAPTER VI

Quantitative Metallography

Most commonly, the examination of a single microstructure is used diagnostically to provide qualitative information on the prior metallurgical history in terms of the number and arrangement of phases. From these, one can deduce the nature and extent of crystallization, transformation, diffusion, and/or deformation processes that have preceded. A number of exemplary microstructures can be used as a set of standards by which some metallurgical feature can be graded as satisfactory or not in terms of performance or manufacturability.

There is growing recognition that systematic functional relationships exist between the dimensions of microstructural features and mechanical or physical properties. This permits quality control of products and processes by the establishment of numerical values of limits applied to one or more microstructural features.

Since most metallurgical materials are polycrystalline, it is not surprising that grain size exerts substantial influence on mechanical properties such as strength (Fig. 6.1). Grain size also correlates with more subtle mechanical property factors such as the ductile–brittle transition (Fig. 6.2). Physical phenomena such as hardenability can be shown to correlate quantitatively with the dimensions of grain size (Fig. 6.3). In these cases, there is a clear implication that is not often stated that the correlations apply to equiaxed grain shapes where a single dimension is sufficient. When the grain shape is elongated, pancake shaped, or ribbon shaped, these correlations are inappropriate. In fact, anisotropy both of grain dimensions and mechanical property performance are to be expected. The choice of plane of polish should reflect the possibility that the grain shape is anisotropic.

Measurement of a metallurgical feature can be used when properly calibrated to provide the specification for a process control. Secondary dendrite arm spacing (Fig. 6.4) correlates precisely with local solidification time. This not only allows regulation of the process by proper

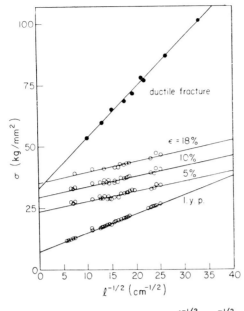

Fig. 6.1. Yield, flow, and fracture stress of mild steel as a function of reciprocal of the square root of the grain diameter. The values of ϵ refer to cold work prestrains; $\bullet$ = data from Patch (1a), $\bigcirc$ = data from Armstrong *et al.* (1b).

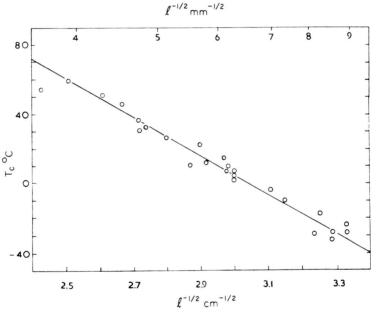

Fig. 6.2. Dependence of the ductile-brittle transition temperature on the logarithm of the reciprocal of the square root of the grain diameter. The material is mild steel (2).

214

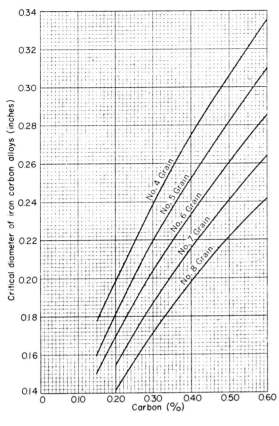

Fig. 6.3. Influence of ASTM grain size number on the critical hardenability diameter of low alloy martensitic carbon steels (3).

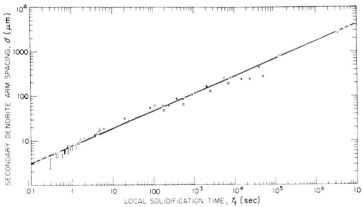

Fig. 6.4 Correlation between measured values of secondary dendrite arm spacing of an Al–4.5% Cu alloy and the measured local solidification times (4).

215

application of chills to a casting, but it provides the possibility of calculation of homogenization times in a subsequent anneal.

Quantitative correlations between distributed phases and mechanical properties have become very important to the reproducible achieve-
ment of high performance. Figure 6.5 demonstrates the powerful
influence of brittle intermetallic compound on the fracture toughness
of a high-strength alloy. The choice of measurement of toughness in
the short transverse direction reflects the recognition of the anisotropy
in this material of both the shape of the compound particles and the
crystals of aluminum alloy solid solution.

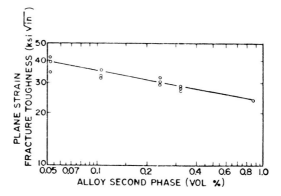

Fig. 6.5. Influence of the volume percent of coarse intermetallic compound
residual from the ingot condition on the plane strain fracture toughness in the
short transverse (thickness) direction. The material is 7075-T6 aluminum
alloy (5).

There are several ways one can quantitatively describe a distributed
phase. Most commonly, volume fraction is used (Figs. 6.5 and 6.6).
Figure 6.6 demonstrates that this is a primary factor but that the identity of the distributed phase must also be a significant factor. Even
crude correlations such as the total number of nonmetallic inclusions
can demonstrate a dominant influence on a mechanical property (Fig.
6.7). More refined correlations would introduce inclusion sizes and
identities. Theoretical considerations suggest correlations that are not
always obvious. Quantitative metallography combined with careful
synthesis of a series of materials with a single variable metallurgical
feature provide important verification (Fig. 6.8).

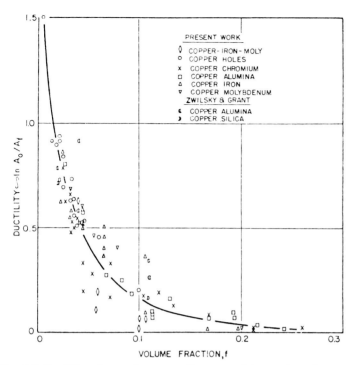

Fig. 6.6 Ductility of several copper alloys containing various dispersions as a function of the volume fraction of the dispersed phase (6).

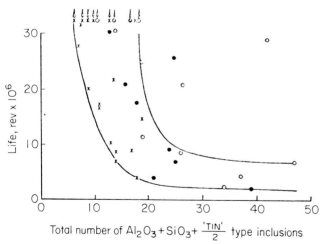

Fig. 6.7. Correlation between total number of nonmetallic inclusions and fatigue life at a fixed stress cycle. The material is a cast steel (7).

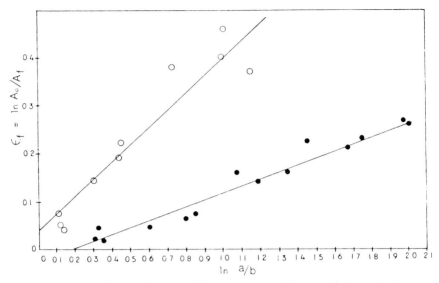

Fig. 6.8. Correlation of measured fracture strain of brass with the ratio of measured pore spacing a to pore diameter b. Open circles represent specimens sintered at 1655°F, solid circles specimens sintered at 1450°F. The material is pressed and sintered 70/30 brass (8).

This chapter concerns itself with procedures by which appropriate measurements of metallographic features can be treated to provide numerical magnitudes of:

the dimensions of distributed phases
the dispersion of distributed phases
the amount of distributed phases
the amount of specific interfacial areas per unit volume
the average volume of grain cells or distributed particles
the distribution of particle sizes

In this subject it is imperative to keep in mind the nature and limitations of the planar section through three-dimensional space which constitutes a microstructure. In a volume occupied randomly by spheres of identical size, any planar section will present circular traces of truncated spheres whose diameters vary over a range from immeasurably small to some maximum value. Thus the diameter of the spheres is not immediately apparent because most of the section traces do not represent the true diameter. Furthermore, it is obviously necessary to take a large number of diameter measurements to find the magnitude of the largest. The determination of the diameter of spheres of equal size randomly distributed in space by observation and measurements in planar section is not difficult largely because of the prior knowledge of uniform size and shape. Nature rarely provides us with such idealized structures. How do we know that the particle size is actually uniform, and if it is not, can average size values and size distributions be estimated? There are parallel problems with particle shapes which are nonspherical. It will be apparent that distributed phases are more often plate- and rod-shaped than spherical.

Quantitative metallography involves large numbers of measurements both because random section traces of solid bodies have a statistical variation and because the size of the solid bodies may not be constant. We are concerned with large numbers of measurements of several kinds on representative planar sections and with the mathematical relationships which permit the detection and extent of real variations, the estimation of good average numerical values of certain useful structural parameters, and the assessment of the accuracy of these numerical values.

Two factors must be assumed or taken into account in all such quantitative studies. It is clearly important that the planar section or sections be representative of the whole. This is an assessment which must be made by inspection. If a systematic nonuniformity in the distribution of phases or in the arrangement of interfaces exists, then, the choice of microscopic fields must compensate. Unfortunately there is no formulated approach to this. There is no good substitute for good judgment.

In the matter of arriving at average values for the dimensions of distributed particles one must have a knowledge of the actual or approximate shape of the particles. Few particles are actually perfectly spherical. Although, it is necessary to treat nodular shapes in terms of spheres of equivalent volume, one should keep in mind that this is a mathematical convenience. Particles crystallizing from either a melt or solid are more often plate- or rodlike. To make the distinction involves rather difficult or tedious work. One can examine the traces of particles near the edge of intersection of two approximately orthogonal polished surfaces. Alternatively one can deeply mark out an area in the field of view under a microscope, for example, with microhardness indentations, and photograph the structure within this periphery at successive layers of repolishing. This is very much like studying the holes in a block of Swiss cheese by examining successive slices. It is possible in some instances by selective dissolution to extract the particles and view them under a stereoptic microscope.

Kinds of Measurements

The quantities which one may wish to determine can be derived from one or more of several kinds of measurements based upon areal analyses, point counting, and lineal analyses.

Areal analysis involves the measurement with a planimeter (or by other means) of the area of a microconstituent intercepted by a planar cross section. Point counting involves the superposition of an appropriate line grid such as transparent graph paper onto a micrograph or real image and for a particular microstructural feature, counting the number per unit area of grid intersections which fall on that type of feature. Lineal analysis involves the estimate of the proportion per unit length of a random line superimposed on a micrograph occupied by a specific type of microstructural feature. In another version of this, and for a different purpose, a count is made of the number of intersections per unit length which a random line makes with a specific type of distributed particle. There are other variations which will be brought up as necessary.

Figure 6.9 diagrammatically illustrates certain characteristics of these systems of measurements. A system of circles of different diameters are randomly disposed on a coarse rectangular grid system. To estimate the volume fracture of the spheres represented by the circles we may compute:

a. $\dfrac{\text{sum of the areas of all of the circles}}{\text{total area of the grid}} = \dfrac{A_\alpha}{A} = 0.139$

b. $\dfrac{\text{number of grid intersections lying in circles}}{\text{total no. of grid intersections}} = \dfrac{N_\alpha}{N} = 0.148$

c. $\dfrac{\text{sum of line lengths lying in circles}}{\text{sum of the lengths of lines 1 to 11}} = \dfrac{L_\alpha}{L} = 0.133$

It is no accident that all of these ratios are nearly identical numerically. A basic rule in quantitative observations made on planar sections is that all three ratios are equal and that in turn they are equal to the volume fraction [see ref. (1) for original references].

$$\frac{V_\alpha}{V} = \frac{A_\alpha}{A} = \frac{N_\alpha}{N} = \frac{L_\alpha}{L}$$

Since each of the three measurement methods is equivalent, one can make a preference based on ease, accuracy, and minimum effort.

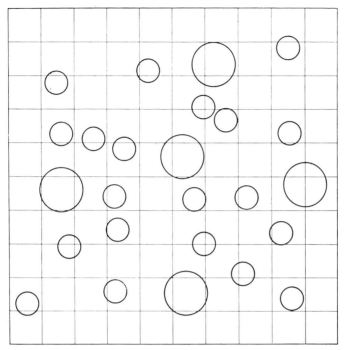

FIG. 6.9. Planar section of spherical particles in a uniform matrix (diagrammatic).

By such consideration, areal analysis can be discounted, for as yet no rapid and convenient method is available. The real choice is between lineal and point count analysis.

Using quantitative measurements of retained austenite in steel as test cases, Howard and Cohen (9) concluded that, whereas both lineal and point counting analysis give reliable results, lineal analysis was preferable from the viewpoints of reproducibility, averaging out segregations, ease, and speed of operation. On the other hand, in answer to the questions:

a. Which procedure is the most efficient in the sense of requiring the least effort for a given precision?
b. Under conditions of maximum efficiency, how many measurements are required to attain a given precision?

Hilliard and Cahn (10) came to the conclusion that the best method for volume fraction analysis is a systematic point count in which a grid is superimposed on a sequence of areas selected either randomly or systematically from the plane of polish. Further, they recommended that to ensure maximum efficiency, it is essential that the magnification relative to the grid spacing should be such that the majority of the structural features do not occupy more than one grid corner. These two preferences based on different criteria of judgment seem to suggest that there is no overwhelming argument for either. It is sufficient to recognize that either method properly used will provide numerical estimates with good precision. Point counting, however, is much less physically exhausting.

Where it is necessary to know the weight fraction of a microconstituent, this can be computed from the volume fraction, the density of the pertinent phase, and the average density of the alloy. The weight fraction of the phase is:

$$\frac{L_\alpha}{L} \cdot \frac{\rho_\alpha}{\bar{\rho}} = \frac{N_\alpha}{N} \cdot \frac{\rho_\alpha}{\bar{\rho}}$$

where ρ is the density of the alloy.

Douglass and Morgan (11) have shown how the solubility of a solid solution in equilibrium with a compound of constant composition may be calculated from the volume fraction of the compound.

$$X = X_0 - (X_c - X_0)\left[\frac{\rho_c}{\rho_\alpha(X)}\right]\left[\frac{f}{1-f}\right]$$

where X = solubility limit of the solid solution, w/o
     X_0 = alloy composition, w/o
     ρ_c = density of the compound
   $\rho_\alpha(X)$ =   density of the solid solution (which mav be dependent on X)
     f = volume fraction of the compound (measured)

Most accurate results are obtained when f is not near its limiting values because small errors in f make for large errors in $f/1 - f$ when $f \rightarrow 1$ and large errors are easy to make when $f \rightarrow 0$.

Measurement of Grain Size

Most of the basic difficulties in quantitative metallography are contained in the problem of assignment of numerical magnitudes to the grains of a polycrystalline aggregate. The traces of grains revealed by a random planar section will not be of equal area nor of identical shape even if in space the grains are of identical shape and size.

Whatever measurement is made, an arithmetic average will be incorrect. It is necessary to take into account the statistics of random sections for contiguous bodies of idealized shape.

From studies of actual grain shapes (12,13) it is clear that while grains do not actually possess regular or idealized forms the truncated octahedron or tetrakaidecahedron is a reasonable approximation. Given a hypothetical aggregate of grains of identical size and of this idealized shape, there are two basic ways of describing it to serve technical scientific purposes. One can designate the number of grains per unit volume or its inverse, the volume of the standard grain. Alternatively one can use some fundamental linear dimension. The volume of a truncated octahedron is equal to $1.414 \, D^3$ where D is the repeat edge dimension indicated in Fig. 1.3. But this dimension is not particularly useful—for example, in relationship between mechanical properties and grain size where the significant parameter is the longest distance of uninterrupted slip within a grain. This is better characterized by the diameter of the inscribed sphere to which the term "grain diameter" properly applies.

One of the basic measurements in grain size determination is the number of grains counted per unit area of planar field of observation. From analysis of the distributions of sections which will be encountered in space occupied by truncated octahedra, the ASTM system gives the following relationship:

$$\log n_A = (N - 1) \log 2$$

where n_A = number of grains per square inch at $\times 100$ magnification and N = ASTM grain size number.

The ASTM grain size number system provides a convenient series of numerical designations for grain size which corresponds roughly to a threefold increment in grains per unit volume for each increment in number. Alternatively the number of observed grains per unit area, n_A, can be converted directly to grains per unit volume n_V as shown in Fig. 6.10 using the data from (14). It can be seen that the two quantities are related by a simple power function:

$$n_V \alpha n_A{}^{3/2}$$

There is an equivalence between the number of grain boundaries intersected per unit length of random line, n_L, and the number of grains per unit area of random section (15):

$$n_V = 0.422 n_L{}^3 = 0.667 n_A{}^{3/2}$$

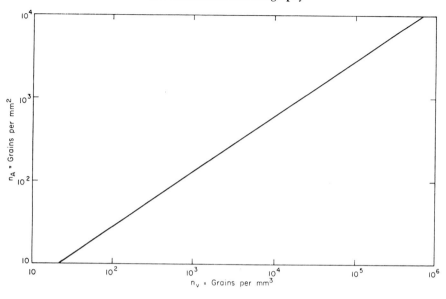

FIG. 6.10. Correlation between planar grain count and volume grain count (*14*).

The grain diameter, d, designating the diameter of an inscribed circle is 1.5 D which leads to the relationship:

$$d = 1.78/n_L = 1.53/n_A^{1/2}$$

This is close to the relationship arrived at by Fullman (*16*) which gives:

$$d = 1.5/n_L = 1.383/n_A^{1/2}$$

The treatment of the subject thus far has assumed that the grains in the aggregate are of uniform size. In any planar section, some small grains will always be seen but if the grain size is truly uniform these will be randomly distributed. If systematic variations in grain. size exist they can to some extent be recognized as clusters of subsize grains. In an analysis of this problem, Hull and Houk (*17*) pointed out that for a uniform grain size, about 95% of the area should be occupied by grains whose diameters are in the range of 2:1. This represents a very tedious process of determination but it can be done if need be. Williams and Smith (*13*) used a method of stereoscopic micro-radiography to study the range of grains sizes in an aluminum-tin alloy wherein the grains of aluminum solid solution are surrounded by a thin film of tin. In this specimen, the grain diameters using visual comparison with standard spheres varied over a range of 8:1 with 74% with a range of 2:1. Probably some modification of the methods used by Brophy and Sinnott (18) and Hyam and Nutting (19) to estimate the distribution of sphere sizes (as described in a subsequent section) would serve also to permit an estimate of grain size distribution.

Implicit in most treatments of grain size is the assumption of the equiaxed shape. If such is not true, as is often the case in practice (see Fig. 1.21), the grain shape anisotropy can only properly be described by grain size measurements in three principal directions.

Other Geometric Properties of Grains

The ratio of grain interfacial area (S) to grain volume (V) is given by Smith and Guttman (20) as follows:

$$\frac{S}{V} = 2N_\mathrm{L}$$

where N_L is the number of intercepts per unit length which a line makes with grain boundaries in planar section. For the case of isolated grains of phase α in a matrix of another phase, the surface to volume ratio is algebraically the same but with a different interpretation of the measured parameters:

$$\left(\frac{S}{V}\right)_\alpha = 2(N_\mathrm{L})_\alpha$$

where (N_L) is a fraction defined as the number of intercepts of grid lines with lines bounding phase α divided by the total length of grid line traversing areas occupied by constituent α.

The ratio of the total length of one-dimensional features to total volume is another way of saying the edge to volume ratio of grains if one were to consider individual grains in space. A one-dimensional feature is a line boundary formed by three adjacent grains. This ratio is given (20) as:

$$\frac{2n}{A}$$

where n is the total number of points (grain corners) seen in area A.

Measurement of Particle Sizes

The volume fraction of a dispersed phase can be determined by simple lineal, areal, or point count analysis and this determination is independent of the shape, ideal or otherwise, which the particles possess. This is not true, however, for the problem of estimating the number of particles per unit volume and the dimensions of particles. In both of these cases, the shape of the particle must be known. In practice one must be able to decide whether the distributed phase approximates a sphere, a disc, a rod, an ellipsoid, or some such ideal geometric shape. Relationships have been derived for particle density and dimensions using parameters measurable on a polished section. Each relationship pertains to a specific geometric shape and assumes a uniform size distribution. The problem of nonuniform size distribution must be considered separately.

Size and Distribution of Spherical Particles

According to Fullman (16) a number of quantities can be determined from two measurements which can be made on planar sections:

N_A = number of particles per unit area
N_L = number of particles per unit length of line

The radius r of spherical particles is defined as follows:

$$r = \frac{2}{\pi} \cdot \frac{N_L}{N_A}$$

The number of spherical particles per unit volume, N_v is given as:

$$N_v = \frac{\pi}{4} \cdot \frac{N_A^2}{N_L}$$

In addition to lineal, areal, and point count analysis, the volume fraction f of a dispersion of spherical particles can be computed as:

$$f = \frac{8}{3\pi} \cdot \frac{N_L^2}{N_A}$$

Size and Distribution of Thin, Circular Plate-shaped Particles

For an ideal circular plate of radius r and thickness t where $t \ll r$, Fullman (16) gives the following useful relationships.

$$r = \frac{N_L}{N_A}$$

$$t = \frac{f}{2N_L}$$

$$N_v = \frac{2}{\pi} \cdot \frac{N_A^2}{N_L}$$

Size and Distribution of Long, Thin Rod-shaped Particles

For an ideal long, thin rod of radius r and length H where $H \gg r$, Fullman (16) gives the following relationships.

$$r = \frac{1}{\pi} \cdot \frac{N_L}{N_A}$$

$$f = \frac{2}{\pi} \cdot \frac{N_L^2}{N_A}$$

Note that, in this case, the length of rods cannot be estimated from measurements made on planar sections.

Size and Distribution of Ellipsoidal Particles

There are two types of ellipsoids—the prolate ellipsoid formed by rotation about the major axes a, and the oblate ellipsoid formed by rotation about the minor axes b. It is necessary to define some parametric terms peculiar to the ellipsoid:

g is the axial ratio b/a for the generating ellipse;

$k_p(g)$ and $k_o(g)$ are nondimensional terms which are functions of g. Numerical values are given for all values of g by DeHoff and Rhines (21).

$\bar{F}_p$ and $\bar{F}_o$ are the average values of the ratio of the minor to the major axis of elliptical traces produced by random planar section through the dispersion of ellipsoids. These would be obtained as the average of a large number of measurements made on a microstructure. Both $\bar{F}_p$ and $\bar{F}_o$ are functions of g, and the graphical relationships are given by DeHoff and Rhines.

$\bar{Z}_p$ and $\bar{Z}_o$ are the average values of the reciprocals of the minor axes of elliptical intersection. These also are derived from measurement of large number of ellipses in the microstructure containing the ellipsoidal particles.

Using these terms, DeHoff and Rhines have defined the number of particles per unit volume and the mean dimensions of the major and minor axes of the ellipsoidal particles:

$$N_v = \frac{2N_A\bar{Z}_p}{\pi k_p(g)} \quad \text{or} \quad \frac{2N_A\bar{Z}_o}{\pi \cdot g \cdot k_o^2(g)} \qquad \bar{b} = \frac{\pi}{2\bar{Z}_p} \quad \text{or} \quad \frac{\pi g^2 k_o(g)}{2\bar{Z}_o}$$

$$\bar{a} = \frac{\pi}{2g \cdot \bar{Z}_p} \quad \text{or} \quad \frac{\pi \cdot g \cdot k_o(g)}{2\bar{Z}_o}$$

Mean Free Path between Particles

The mean free path is the average distance between the peripheries of particles in space. This distance is independent of particle shape for a particle dispersion, and is given by Fullman (16) as:

$$\frac{1 - f}{N_{\mathrm{L}}}$$

Particle Size Distribution Functions

A form of lineal analysis can be used to arrive at the distribution of particle sizes of certain ideal shapes. This work involves the meas-

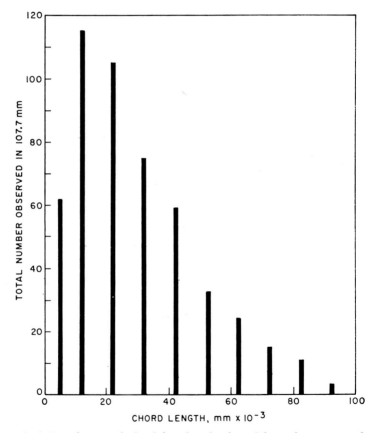

FIG. 6.11. Distribution of chord lengths of spheroidal graphite measured on a polished section of nodular cast iron. After Brophy and Sinnott (18).

urement of the various intercept lengths encountered along an arbitrary line or number·of randomly chosen lines. With sufficient number of measurements one can graphically represent the population of intercept lengths in the planar field. Figure 6.11 presents such data for the population of chord lengths in the microstructure of a cast iron containing a dispersion of spherical graphite particles which in planar section appear as circles. The same sort of measurement can be applied to the thickness of platelets in a eutectoid structure.

The form of the population curve itself is an indication of whether or not the particle size is uniform. If one considers the population of space by spheres, the form of the chord population can be appreciated intuitively. For a uniform sphere size, the magnitude of any chord encountered by a random line cannot exceed the diameter of the sphere. The population curve of chords must end abruptly at some maximum value giving a sawtooth curve form. With the existence of two or three sphere sizes, the population curve must be the superposition of sawteeth subdistribution. Finally if a continuous and systematic distribution of sphere sizes exist, the population of chords can lose any characteristic shape. These points are illustrated in Fig. 6.12.

For the distribution of sphere sizes in an opaque body both Cahn and Fullman (22) and Brophy and Sinnott (18) give the same treatment.

$$N(D)_{D=t} = -\frac{2}{\pi} \cdot \frac{d[n(t)/t]}{dt}$$
$$= \frac{2n(t)}{\pi t^2} - \frac{2}{\pi t} \cdot \frac{dn(t)}{dt}$$

where $N(D)_{D=t}$ is the number of spheres whose diameter lie between t and $t + \Delta t$.

$n(t)$ is the number of chords per unit length of line traverse whose lengths lie between t and $t + \Delta t$. This is the measurement made in lineal analysis.

This equation can be solved graphically or numerically. The numerical approach seems simplest. The sphere size distribution function rewritten for numerical analysis is:

$$N(D)_{D=t} = -\frac{2}{\pi} \cdot \frac{d[n(t)/t]}{dt}$$
$$= \frac{2}{\pi} \left[\frac{\dfrac{n(t)}{t} - \dfrac{n(t + \Delta t)}{t + Dt}}{\Delta t} \right]$$

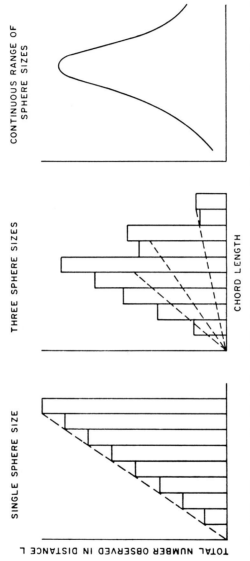

FIG. 6.12. Diagrammatic illustration of the influence of the number of sphere sizes on the distribution of chords in planar section revealed by lineal analysis.

In Table 6.1, a portion of Brophy and Sinnott's data and calculations are reproduced with some changes in notation to conform to those used here. The data relate to the distribution of chord lengths of the traces of graphite spheres in a nodular cast iron.

Table 6.1

SAMPLE CALCULATED SPHERE SIZE DISTRIBUTION
IN A SPECIMEN OF NODULAR IRON[a]

Chord length interval (mm × 10³)	No. of chords counted in interval[b] $n(t)_c$	Medium chord length $\bar{l}$ (mm × 10³)	$\dfrac{n(t)_c}{L\bar{l}\Delta t} = \dfrac{n(t)}{\bar{l}\Delta t}$	$\dfrac{n(t)_c}{L\bar{l}\Delta t} - \dfrac{n(t+\Delta t)_c}{L(t+\Delta t)\Delta t}$	No. of spheres	Diameter of spheres (mm × 10³)
2.5–7.5	64	5	23,800			
				15,260	9710	7.5
7.5–17.5	115	12.5	8,540			
				4200	2673	17.5
17.5–27.5	105	22.5	4,340			
				2195	1344	27.5
27.5–37.5	75	32.5	2,145			
				855	545	37.5
37.5–47.5	59	42.5	1,290			
				707	450	47.5
47.5–57.5	33	52.5	583			
				226	144	57.5
57.5–67.5	24	62.5	357			
				165	105	67.5
67.5–77.5	15	72.5	192			
				63	43.2	77.5
77.5–87.5	11	82.5	124			
				103.9	66.1	87.5
87.5–97.5	2	92.5	20.1			
				20.1	12.8	97.5

[a] After Brophy and Sinnott (18).
[b] Over total traverse length $L = 107.7$ mm. $n(t) = [n(t)_c]/L$.

Another method which seems simple to perform was outlined by Hyam and Nutting (19). For a random planar section through a body populated by spheres of a constant diameter, the probability of finding circles of diameter between zero and any diameter, d, which is less than the diameter of the spheres is given by:

$$P_D(d) = 1 - \frac{1}{D}\sqrt{D^2 - d^2}$$

where D is the diameter of the spheres and $D > d > 0$. If the count of circle diameters per unit area of observation is arranged in a $\sqrt{2}$ progression of numbers, the numerical magnitudes of $P_D(d)$ become independent of D. This is to say that d assumes values $D/\sqrt{2}$, $D/\sqrt{2^2}$, $D/\sqrt{2^3}$, $D/\sqrt{2^4}$, and so on. The numerical values of $P_D(d)$ are given in Table 6.2.

Table 6.2

PROBABILITIES OF OBTAINING CIRCLES IN STATED SIZE
RANGE FROM SECTIONING SPHERES OF DIAMETER D[a]

Circle diameter size group limits	Probability $P_D(d)$
$D - D/\sqrt{2}$	0.707
$D/\sqrt{2} - D/\sqrt{2^2}$	0.159
$D/\sqrt{2^2} - D/\sqrt{2^3}$	0.069
$D/\sqrt{2^3} - D/\sqrt{2^4}$	0.033
$D/\sqrt{2^4} - D/\sqrt{2^5}$	0.016
$D/\sqrt{2^5} - D/\sqrt{2^6}$	0.008
$D/\sqrt{2^6} - D/\sqrt{2^7}$	0.004
$D/\sqrt{2^7} - D/\sqrt{2^8}$	0.022

$N_c = N_s \times P_D(d)$
N_c = number of circles in unit area of planar section.
N_s = number of spheres in unit volume.

[a] After Hyam and Nutting (19).

Consider now the condition where the sphere diameters are variable. Figure 6.13 presents a hypothetical distribution of circle diameters arranged so that each successive circle diameter grouping differs from the next according to the $\sqrt{2}$ progression. Group 6 therefore represents the number of circles counted in unit area of field whose diameters vary from the largest, d_{max}, to $d_{max}/\sqrt{2}$. We may assume that all circles in this group 6 represent true diametral sections of the sphere distribution. Using Table 6.2, the number of spheres in this size group, N_s, is computed from the number of circles, N_c:

$$N_s(6) = N_c(6)/0.707$$

The circles in group 5 represent some diametral planar sections of spheres and some under diametral planar sections from the spheres of group 6. From Table 6.2, the number of circles contributed by $N_s(6)$ to those found in group 5 is:

$$N_c{}^6(5) = N_s(6) \times 0.159$$

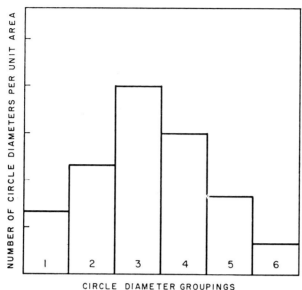

CIRCLE DIAMETER GROUPINGS

Fig.6.13. Hypothetical histogram of circle diameter frequency observed on a plane of section. Each group limit differs from the adjacent one by a factor of $\sqrt{2}$.

By difference, $N_c(5)$ measured $- N_c{}^6(5)$ gives the circles which are true diametral planes, whence:

$$N_s(5) = N_c(5) \text{ meas.} - \dot{N}_c{}^6(5)\ 0.707$$

Similarly $N_s(4) = [N_c(4) \text{ meas.} - N_c{}^6(4) - N_c{}^5(4)]/0.707$

Although simple enough computationally, this method permits an error in counting to carry through the whole calculation. For instance, an error in $N_c(6)$ meas. would have a large effect on $N_s(5)$, a small effect on $N_s(4)$, and a very small effect on $N_s(3)$.

For the distribution of plate thicknesses a treatment analogous to the problem of spheres is given by Cahn and Fullman (22).

$$V(S)_{s=1} = 3.1\ m(l) + l^2\ \frac{dm(l)}{dl}$$

where $V(S)_{s=1}$ is the volume fraction occupied by platelets with thickness S

l is any intercept length in line traverse across a structure

$m(l)$ is the number of intercepts per unit length of line traverse having an intercept length between l and $l + \Delta l$.

Measurement of Dihedral Angles

The original shape of particles of a distributed phase is largely governed by factors which maximize the rate of growth. After the growth process has ceased, the particle will continue to change shape under the influence of interfacial tensions. When the particle lies within the interior of a grain, the interface is a continuous uninterrupted surface with the surface tension vectors of almost constant magnitude lying in this surface. The condition of minimum interfacial

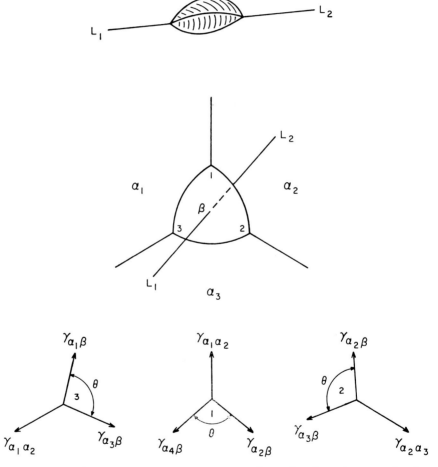

FIG.6.14. Shape, transverse cross section, and interfacial tensions associated with a β particle lying on the line of intersection of three α grains.

energy corresponds to the condition of minimum surface to volume, i.e., a spherical shape. The mechanism by which the spherical shape is approached may involve creep deformation under the action of the envelope of tension or surface diffusion between points of different radius of curvature. The problem of which mechanism is dominant is the same as for the change of shape of voids during sintering, and the dilemma is not well resolved in either case.

When the particle lies in the plane of abutment between two grains of the majority phase, or along the line of juncture of three grains or at the point of mutual contact of four grains, the equilibrium shape will be distorted from the ideal sphere because of the resultant of surface tension vectors acting along specific directions. Consider the cross section of a particle of minority phase lying in the line of intersection, L_1L_2, between three grains, α_1, α_2, α_3, illustrated in Fig. 6.14. At each of three points a triangle of interfacial tensions exists which distorts the ideal spherical (circular trace) form. In general the differences in magnitudes among $\gamma_{\alpha_1\alpha_2}$, $\gamma_{\alpha_2\alpha_3}$, $\gamma_{\alpha_1\alpha_3}$ are small compared with the differences between $\gamma_{\alpha\alpha}$ and $\gamma_{\alpha\beta}$. One can therefore write a tension balance:

$$\gamma_{\alpha\alpha} = 2\gamma_{\alpha\beta} \cdot \cos\frac{\theta}{2}$$

where θ is the angle contained by the $\gamma_{\alpha\beta}$ vectors and is called the dihedral angle. In the microstructure this angle is contained by the tangents to the α/β boundaries at the $\alpha/\alpha/\beta$ juncture. The degree of distortion from the spherical shape depends on the ratio $\gamma_{\alpha\beta}/\gamma_{\alpha\alpha}$ as illustrated in Fig. 6.15. Thus for $\gamma_{\alpha\beta}/\gamma_{\alpha\alpha} \leq \frac{1}{2}$, $\theta = 0$, i.e., the grains of α become enveloped by β; whereas when $\gamma_{\alpha\alpha} \ll \gamma_{\alpha\beta}$, $\theta \to 180°$, i.e., the undistorted sphere.

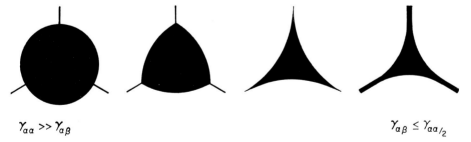

$\gamma_{\alpha\alpha} \gg \gamma_{\alpha\beta}$ $\gamma_{\alpha\beta} \leq \gamma_{\alpha\alpha/2}$

FIG. 6.15. Influence of relative interfacial energy on the shape of second phase (β) particles lying on the line of intersection of three matrix (α) grains.

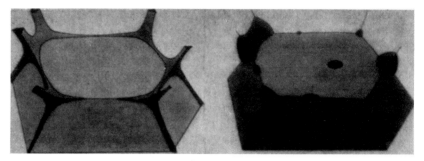

FIG.6.16 Sketch showing distribution of second phase in three dimensions. Left
θ = about 65°; Right θ = about 120° Drawing by C. S. Barrett. Taken from
Smith (23).

The actual spatial shapes that interfacial tensions will impose on
particles lying between grains of the majority phase can be appreciated
from the sketches reproduced in Fig. 6.16 from the basic paper on this
subject written by C. S. Smith (23). Because of the significant influence
of relative interfacial energies on the shape and disposition of minority
phases in a microstructure, the measurement of dihedral angles
provides important information to the physical metallurgist. It must
be emphasized that the dihedral angles permit the estimation of
relative interfacial energies only. However, by using multiphase systems it is possible to use the few absolute magnitudes available to
compute new absolute magnitudes for various interfaces [see for
example Van Vlack (24)].

It is obvious that the angles measured in random planar sections of
a single dihedral angle cannot be all the same. Smith (23) has shown
that the most frequently observed dihedral angle in a planar section
is, in fact, the true spatial dihedral angle. Exemplary results of an
analysis of the dihedral angles in an alloy of copper and 37% zinc
containing α phase particles in a polycrystalline aggregate of β phase
grains are shown in Fig. 6.17. The most frequently occurring angle
between intersecting α/β interfaces is 100° which is taken as the true
dihedral angle and from which it may be estimated that $\gamma_{\alpha\beta}/\gamma_{\alpha\alpha} = 0.78$.
The same type of analysis can be applied to the groove angles which
develop in the grain boundaries which emerge to the surface of a
single phase specimen immersed at elevated temperatures in a liquid
or gaseous medium. The number of measurements which have to be
made is no more than is needed to establish a continuous distribution
and a clear preference for one angular value.

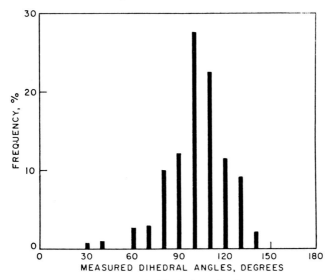

FIG.6.17. Distribution of observed dihedral angles between α/β interfaces in an $(\alpha + \beta)$ brass containing 37% Zn. From Smith (23).

Riegger and Van Vlack (25) have shown that greater accuracy with fewer measurements can be obtained by plotting cumulative per cent of measured angles against a scale of angles. The true dihedral angle is given by the distribution curve at 50 cumulative per cent. The authors claim good accuracy with only 25 measurements.

Accuracy in Quantitative Analysis

Under the best of circumstances, quantitative metallography is a tedious procedure. Clearly, therefore, the extent of effort should be no more than is necessary. The question of how many measurements to make is answered in terms of the desired accuracy. In statistics there is a standard way of describing accuracy. For this purpose, we introduce some appropriate nomenclature and symbols. In the determination of proportion of a phase, let the true proportion be p. With as many measurements as are taken, a best estimate of p is given as Z. In analyzing all of the data, the final value of Z contains a standard deviation σ. Suppose that the best estimate Z of the volume proportion of a phase is 0.50 and we wish that this numerical value were accurate to within ± 0.05. Whether it be ± 0.005 or ± 0.10 or ± 0.001 is a matter of judgment of the real need; for with a demand for high accuracy will come a proportionate demand of effort. It is not possible to be certain of 0.50 ± 0.05 but it is possible to attain a 95% or 99% probability or confidence. According to Gladman and Woodhead (26) in point counting, provided that the total number of points counted exceeds about 200 and that $0.9 > p > 0.1$, the 95% confidence limits would allow the $\pm 0.05\%$ of our hypothetical situation to be equated to $\pm 1.96\sigma$ and the 99% confidence limits to provide that $\pm 0.05 = \pm 2.6\sigma$. In other words statistics provide that the true value of p has a 0.95 probability of being between $Z - 1.96\sigma$ and $Z + 1.96\sigma$.

The specific need for an accuracy of 0.50 ± 0.05 has now been transposed to a definition of the magnitude of the standard deviation σ, i.e. $\sigma = 0.026$ (95% confidence). Having made an arbitrary number of measurements, one can analyze the data and determine whether enough has been done to meet the requirement. Some further definitions are in order. The point counting method is used as a basis of illustration.

A number, k, of groups of measurements, each involving n points are made. For each group of n points, an estimate Z ($= n\alpha/n$) of p can be made. Thus for k groups, k values of Z are derived. The best estimate of p is the mean value of Z.

$$\bar{Z} = \frac{\Sigma Z}{k}$$

For each group of n points, the standard deviation for the Z derived is:

$$\sqrt{\frac{(Z - \bar{Z})^2}{n - 1}}$$

For the total number of measurements, nk, the standard deviation or standard error is:

$$\sigma = \sqrt{\frac{(Z - \bar{Z})^2}{k(k - 1)}}$$

The relationship between true proportion of a phase, the standard deviation and the total number of points used in point counting analysis has been developed by Gladman and Woodhead (18).

$$N = \frac{p(1 - p)}{\sigma^2}$$

$N = nk$ in the previous notation.

The validity of this relationship is predicated on certain restrictions about the conditions of measurements. It is required that the constituents in the microstructure be distributed at random, that the number of particles shall be large compared to the number of grid points, and that the structure shall be finely divided relative to the grid spacing.

From a visual estimate of p and a decision as to the desired magnitude of σ, one can arrive at a minimum number of points to be used in the count. Values of N, p, and σ are given in Table 5.3. Having

Table 6.3 [a]

NUMBER OF POINTS, N, AS A FUNCTION OF p AND σ

	σ					
p	0.001	0.0025	0.005	0.0075	0.010	0.025
0.02	19600	3163	784	348	196	
0.04		6144	1536	683	384	
0.06		9024	2256	1003	564	
0.08			2944	1308	736	
0.10			3600	1600	900	144
0.15			5100	2267	1275	204
0.20			6400	2844	1600	256
0.25				3333	1875	300
0.30				3733	2100	336
0.40					2400	384
0.50					2500	400

[a] After Gladman and Woodhead (26).

made the N counts it will be found that the original estimate was somewhat in error and that the original goal of N counts was either inadequate or more than adequate. At any stage in the analysis, the existing data can be reduced to its value of σ to check whether the job has gone far enough. Some of the data from the Gladman and Woodhead paper are reproduced in Table 6.4. The data relate to the point count analysis of the volume fraction of pearlite in a normalized 0.29% carbon steel. From all of the measurements made, the volume fraction of pearlite is 0.556. The measurements were made in groups and various aggregates of these are analyzed to provide a standard deviation based on the data and as-calculated from the relationship above for $N = fcn\ (p,\sigma)$.

Hilliard and Cahn (10) derive a somewhat different expression:

$$N = \frac{P^2}{\sigma^2}$$

Table 6.4[a]
POINT COUNT ANALYSIS OF PEARLITE IN A NORMALIZED STEEL ($p = 0.556$)

No. of points in group, n	No. of groups, k	Total no. of points, nk	Mean proportion, $\bar{Z}$	Standard deviation from: Data	Standard deviation from: Eqs.
25	20	500	0.572	0.0195	0.0222
25	20	500	0.526	0.0220	0.0222
25	40[b]	1000	0.549	0.0150	0.0157
50	15	750	0.555	0.0293	0.0181
120	5	600	0.573	0.0233	0.0203
10	25	250	0.548	0.0361	0.0314

[a] After Gladman and Woodhead (26).
[b] These 40 groups are a combination of the first two sets of 20.

but, in a footnote, they recommend the use of the Gladman-Woodhead relation. Equivalent relations for the cases of areal analysis, lineal analysis, and random point counting are derived also, but it is apparent that their use to check whether an adequate number of measurements has been made involves considerably more work. For example the standard deviation for a lineal analysis where the lineal fraction is small is:

$$\frac{\sigma}{P^2} = \frac{(\sigma_t/\bar{t})_\alpha{}^2 + 1}{1/N_\alpha}$$

where $\bar{t}$ = mean intercept length of α features

σ_t = standard deviation of the intercept lengths of α features

$N\alpha$ = total number of intercepts encountered

It is necessary first to derive the magnitude of $(\sigma_t/\bar{t})$ which is a dimensionless measure of the degree of dispersion in the size distribution of α features on the plane of polish. Only then can the volume fraction determination be measured for adequacy in the number of measurements.

This subject is much too complex to be dealt with adequately in a small part of a small book. The chapter has been written to draw attention to the needs for quantitative estimates of metallographic details in physical metallurgy and to introduce some of the elementary considerations involved in the acquisition of measurements and their use in estimations. For a comprehensive treatment of the subject, the reader is referred to references (27) and (28). Another review chapter is given in references (29) and (30).

REFERENCES

1a. N. J. Petch, *J. Iron Steel Inst.* (*London*) **173**, 25 (1953).
1b. R. W. Armstrong, I. Codd, R. M. Douthwaite, and N. J. Petch, *Phil. Mag.* **7**, 45 (1962).
2. N. J. Petch, "The ductile-cleavage transition in alpha-iron" *in Conference on Fracture—Swampscott, N.A.S., 1959*, pp. 2.1–2.15.
3. "Modern Steels and their Properties" (6th ed.). Bethlehem Steel Publication.
4. T. F. Bower, H. D. Brody, and M. C. Flemings, *Trans. AIME* **236**, 624 (1966).
5. J. H. Mulherin and H. Rosenthal, *Met. Trans.* **2**, 427–432 (1971).
6. B. I. Edelman and W. M. Baldwin, *Trans. ASM* **55**, 230–250 (1962).
7. J. D. Murray and R. F. Johnson, "Clean Steel," British Iron and Steel Institute Publ. #77, 110–118 (1963).
8. W. Rostoker and S. Y. K. Liu, *J. Mater.* **5**, 605–617 (1970).
9. R. T. Howard and M. Cohen, *Trans. AIME* **172**, 413–426 (1947).
10. J. E. Hilliard and J. W. Cahn, *Trans. AIME* **221**, 344–352 (1961).
11. D. L. Douglass and R. E. Morgan, *Trans. AIME* **215**, 869–870 (1959).
12. C. H. Desch, *J. Inst. Metals* **22**, 241, 263 (1919).
13. W. M. Williams and C. S. Smith, *Trans. AIME* **194**, 755–765 (1952).
14. "Metals Handbook" (1948 ed.), p. 405. ASM, Metals Park, Ohio.
15. E. E. Underwood, *ASM Eng. Metals Quart.* **1**, 70–81 (1961).
16. R. L. Fullman, *Trans. AIME* **197**, 447–452 (1953).
17. F. C. Hull and W. J. Houk, *Trans. AIME* **197**, 565–572 (1953).
18. J. H. Brophy and M. J. Sinnott, *Trans. ASM* **54**, 65–71 (1961).
19. E. D. Hyam and J. Nutting, *J. Iron Steel Inst.* (*London*) **184**, 148–165 (1956).
20. C. S. Smith and L. Guttman, *Trans. AIME* **107**, 81–87 (1953).
21. R. T. DeHoff and F. N. Rhines, *Trans. AIME* **221**, 975–982 (1961).
22. J. W. Cahn and R. L. Fullman, *Trans. AIME* **206**, 610–612 (1956).
23. C. S. Smith, *Trans. AIME* **175**, 15–51 (1948).
24. L. H. Van Vlack, *Trans. AIME* **191**, 251–259 (1951).
25. O. K. Riegger and L. H. Van Vlack, *Trans. AIME* **218**, 933–935 (1960).
26. T. Gladman and J. H. Woodhead, *J. Iron Steel Inst.* (*London*) **194**, 189, (1960).
27. R. T. Dehoff and F. N. Rhines (eds.), "Quantitative Metallography." McGraw-Hill, New York, 1968.
28. E. E. Underwood, "Quantitative Stereology." Addison-Wesley, Reading, Massachusetts 1970.
29. E. E. Underwood, Applications of quantitative metallography, *in* "Metals Handbook," (8th ed.) Vol. 8. ASM, Metals Park, Ohio, 1973.
30. H. E. Exner, Review 159, *Metallurg. Rev.* **17**, 25–42, 1972.

CHAPTER VII

Energy Dispersive Spectrography (EDS)

Introduction

X-Ray Detection

When atoms in a sample are bombarded by photons (x rays or γ rays) or high-velocity particles (electrons, protons, α or β particles) energy is transferred to the atoms by exciting electrons in the atom. The excess energy released during movements of these electrons is expressed as an x-ray photon. The energy of the photon is the same as the difference in energy levels of the atom and can be used to identify the atom. Analysis is performed by collecting and sorting the emitted x radiation. Using these methods, elements with atomic number 11 and above can be identified. Equipment now available can detect elements down to atomic number 5. Data are accumulated as a spectrum of the distribution of the x-ray wavelengths detected versus their energy or intensity. The energy of an emitted x-ray wavelength is a measure of the amount of the element present.

The generation of x rays in the scanning electron microscope (SEM) is a somewhat inefficient process. Generally, one characteristic x-ray photon is emitted per thousand electrons striking a sample. The collecting, sorting, and quantifying of emitted x rays can be quite complex. EDS spectrometers tend to suffer from lack of energy resolution. The long dead time of an EDS spectrometer along with its low peak-to-background ratio results in what are considered poor detection limits. In general, it should be expected that analyses will be confined to a minimum of 0.5 wt. % of the total unknown. It is thought that, in many cases, analyses showing less than 1 wt. % are questionable. The area of x-ray bombardment is in reality a teardrop-shaped mass that penetrates the sample to a depth of a few micrometers to a millimeter, depending on the x-ray energy used and the matrix composition of the sample. EDS analysis is not truly a surface analysis. In many instances it is possible to penetrate a small inclusion completely and excite it as well as matrix

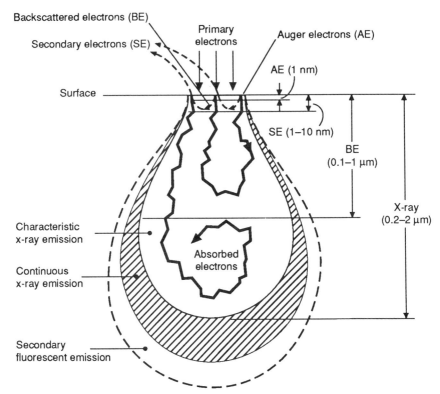

FIG. 7.1. Types of electron beam-excited electrons and radiation used in SEM and depth of the region below the specimen surface from which information is obtained. From H. E. Exner, Scanning electron microscopy, *in* "Metals Handbook" (9th ed.), Vol. 9, p. 90. American Society for Metals, Metals Park, Ohio, 1985.

material surrounding it. See Fig. 7.1 for a diagrammatic illustration of the excitation zone.

The electron beam size in an SEM is 0.1 to 1.0 μm in diameter and, as can be seen in Fig. 7.1, a significantly larger volume of the sample is excited. Particles to be analyzed must be larger in volume than the excited portion. One can obtain a "bulk" analysis by scanning the sample at a relatively low magnification. Composition differences on a small scale can play an important part in the perceived bulk analysis. As a result, many areas should be examined and the analytical results averaged to obtain reliable numbers.

X-Ray Analysis

The most accurate EDS quantitative analyses are based on references to known standards. Comparison to a standard with an exact composition match is ideal but most often unfeasible. The accepted method is to compare to standards with close to known amounts of the element of interest. Pure elements are frequently used. EDS software compares the standard to the unknown and calculates a composition. In many instances elemental peaks overlap, which makes analysis difficult. Again, the software sorts and calculates as well as it can. In these instances one must rely on secondary and tertiary element peaks to properly define the element content. It is important then to standardize the EDS system accurately and check it at frequent intervals. For example, the sulfur $K_{\alpha 1}$ peak is present at 2.308 keV and the $K_{\alpha 2}$ peak is at 2.306 keV. But the molybdenum $L_{\alpha 1}$ peak is at 2.293 keV and the $L_{\alpha 2}$ peak at 2.290 keV. The peaks for these two elements are too close, and if any instrument instability or drift occurs, they cannot be correctly distinguished.

Sample Preparation

SEM image contrast is obtained by emission of electrons from edges, corners, and surface protrusions. With a flat sample, such as a metallographically polished specimen, the contrast and image quality can be poor. It can be difficult to locate a specific region of interest as identified by light optical microscopy. It is possible to use microhardness indentations to outline or point to a particle or field of interest. It is also possible to scribe a mark on the polished surface to locate an area if it is large enough to see at low magnification.

Generally, a sample must be electrically conductive to be viewed in the SEM. A path to ground must be provided. An ungrounded sample charges up under impingement of the electron beam. Image quality and beam stability suffer, and it can be quite difficult to obtain an analysis of a small particle. It is a simple technique to coat a mounted metallographic specimen with carbon by using vacuum evaporation techniques. The presence of carbon provides a conductive film and does not interfere with EDS analysis by most systems used today. Better results can be obtained by coating with a metal such as gold. Metal coatings are detected and must, of course, be excluded from the analysis. Newer thin-window or windowless detectors, however, can see lighter elements including carbon. Conductive coatings range in thickness from 50 to 500 Å. Conductive mounting materials are also available and their use can be of help in any SEM examination.

Sample flatness is very important. In single-phase, homogeneous samples, sample preparation is relatively simple. When phases of different hardness coexist, steps may be present at the phase boundaries. These must be taken into account if EDS analysis is to be performed near the boundary. One way is to rotate the sample 180° and remeasure it. Just the surface relief produced between grains of single-phase material can also cause problems. It is also possible with improper polishing to embed polishing abrasive in the sample surface. "Inclusions" containing the elements that are present in the polishing abrasive should be viewed with caution. Poor polishing techniques can cause surface smearing, which can also alter the analytic results.

Forensics

Although forensics is commonly understood as the search for specific details on specimen materials of importance in litigation, the engineer or researcher is also called on to provide information not involved in legal matters but on a specimen material consisting of little more than a fragment. Of course, when larger quantities are available, simpler and more extensive methods of examination can be used than EDS, but the emphasis here is on situations where only small fragments have been recovered. We assume that the fragment is large enough to be mounted for metallographic preparation. It might be no larger than a pinhead. The smallness may not just refer to a discrete fragment. The specimen could be a thin layer as on a brazed joint or an electroplated coating.

When the specimen has been mounted and polished, its identity in broad terms should be determined, if for no other reason than to choose a proper etchant. The EDS energy profile will identify not only the major elements but also many of the minor ones. Thus one can identify the base metal and make a good guess at the the alloy category. This will be quite sufficient in many cases. Finer details require quantification and can be obtained within the limits of the EDS system.

With this preliminary information, the microstructure can provide definition of the state of manufacture in terms of cast or wrought state and heat treatment. Microhardness is helpful in making certain distinctions.

Figure 7.2 shows a few examples of the simplest preliminary EDS studies, and in each case the legend provides a brief comment on where that leads.

Phase Identification

A considerable part of the objectives of metallography is to get a first look at the entities (phases) that make up the body of a solid. There is generally enough preliminary information to provide some expectations, but the information may be erroneous or deficient. These explorations often lead to more questions than answers.

Commonly, the expectation is a single- or two-phase structure that would be characteristic of the general information about composition. But the metallographic examination shows more phases than expected or a configuration of phases that is not consistent with what is thought to be the process history. The compositional identity of discrete phases or of contrast etching zones can go a long way toward providing answers to the questions raised. In the practice of physical metallurgy, early diagnosis leads to interpretation or a suggested hypothesis.

But an interpretation or hypothesis is rarely more than a beginning. Validation gives credibility to what might otherwise be glib conjecture. The early metallography and EDS analyses of phases provide the frame of reference between initial insights and attempts to reproduce the material or set it against existing standards.

We demonstrate such situations by three cases of process development. In the case illustrated in Fig. 7.3, a toxic entity is to be isolated from the environment in a cost-effective manner. The material is not a metal but a slag which might be broadly termed a ceramic material. In the second case, presented in Fig. 7.4, a coating is being developed to defend against environmental attack. The concept seems sound and the resulting performance looks good, but what do we have? Are there defects in the coating that might result in failure in an industrial setting? In the third case (Fig. 7.5), we have a by-product from secondary smelting which is, by early general analysis, a ferronickel with marketable value as an addition or as an engineering material on its own.

Composition Gradients

A significant part of metallurgy has to do with modifications of the surfaces of materials. There are processes that are used to enrich the surface with one or more elements in order to improve the surface appearance or performance. In service, conditions may exist where unwanted depletion or enrichment of alloying elements occurs in the surface of a material. Simple observation of color or roughness may suggest these occurrences; and cross-section metallography may indicate them. However, details of depth, the elements involved, and actual

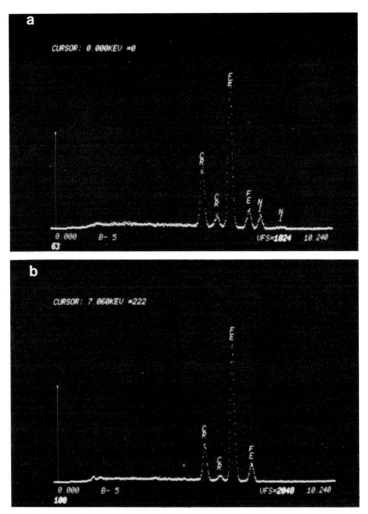

Fig. 7.2. Each of the samples represented here was mounted and polished in the conventional manner of metallographic preparation. We present the energy spectrum and the elements which they represent. Note that elements often have more than one characteristic energy, and two elements can have characteristic energies so close that their identity is ambiguous. Tables of characteristic wavelengths of the elements indicate such interferences.

Samples (a) and (b) are two stainless steels, an austenitic grade and a ferritic grade. Specimen (a) is identified as the austenitic grade because of the joint presence of chromium and nickel. Within the 300 series of stainless steels, further discrimination is difficult except where, as with the 316 grades, the additional presence of molybdenum can be noted. The ferritic grade (b) is distinguished by the presence of chromium only. Of course, other signs can be used as well. The presence of annealing twins characterizes the austenitic grades and simple ferromagnetic indications characterize the ferritic grades.

248

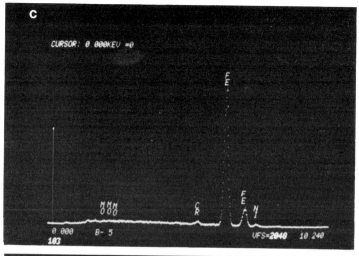

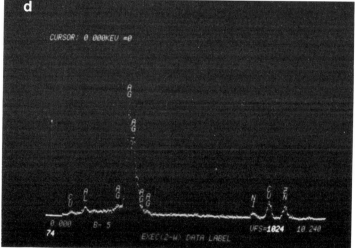

Sample (c) is another steel with small amounts of nickel, chromium, and molybdenum. The evidently small amounts indicate a heat-treatable, alloy carbon steel. The AISI or SAE 43xx and 86xx series possess this combination of alloying elements. Quantification of the relative proportions of nickel and chromium can make a further distinction. Metallographic measures of ferrite/pearlite proportion can provide further evidence of the carbon content, which is not within the scope of EDS.

Sample (d) is the energy spectrum of a brazing alloy. The major element is silver and the alloy additions are copper, zinc, and nickel. Among the silver solders, this combination of alloying elements seems to narrow the identification to the grade designated as BAg-4. If the brazing alloy is actually part of a brazed joint, the compositions may have been modified by solution and diffusion accompanying the brazing operation, as it involves interaction with the massive pieces being joined.

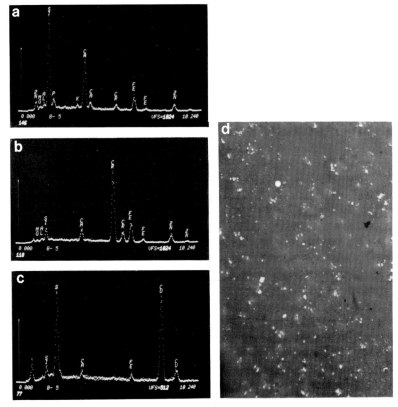

FIG. 7.3. Spent electroplating solutions and rinse waters contain appreciable amounts of metallic ions in solution which are toxic even in very low concentrations. Other ions are noxious or hazardous to plant life. It is increasingly demanded that water be returned to the environment as clean as it was when taken for use. The established procedure is to precipitate metallic ions as hydroxides by adjusting the water to a high pH, flocculating, and filtering. The extracted residue is a filter cake which can be dried but is unsafe because the hydroxides are vulnerable to slow re-solution in the presence of ground waters.

A process has been devised to create a fused product in which the oxides of the toxic metals are locked into a solid substance from which they do not leach at any significantly hazardous rate. Standards of leach resistance have been set and are continually revised to match improvements in waste treatment.

The example here is a slag produced by fusion with appropriate additives of a filter cake containing a considerable amount of chromium in oxide form. The unetched micrograph (d) of the slag (×650) shows a matrix populated by a dispersion of fine particles, one of which is characteristically spheroidal. That usually means that immiscibility of liquids developed in the fusion or in early stages of cooling. The angular shape shows that the majority dispersed phase is a primary crystallization product. Beside the micrograph, the EDS traces (a–c) of secondary energy emissions and the elements they signify for each of the three phases are shown.

The matrix phase is a calcium ferrosilicate, which is known to be very leach resistant. The majority dispersed phase is predominately a chromium oxide that has been isolated from water by the matrix. The globular phase is a copper sulfide. Note that water-soluble salts have been occluded with the filter cake, and these must have been sulfates and phosphates. Also, besides chromium and copper plating, zinc plating solutions have been added for treatment at the same time.

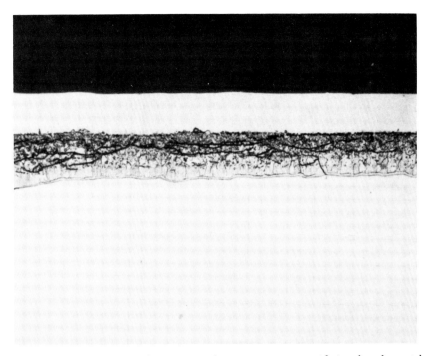

Fig. 7.4. The silicides of titanium are far more resistant to oxidation than the metal itself, and a coating of such silicides could substantially raise the operational temperatures of titanium and its alloys in service. The micrograph ($\times 400$) is an example of an experimental approach to producing such a coating. The edge of the coating has been protected from spalling off during cutting, mounting, grinding, and polishing the sample for examination by deposition of a thin layer of electroless nickel, which is very hard. The coating is above the silicide layer in the micrograph.

The specimen is seen to have at least three layers, one very thin at the titanium alloy substrate boundary. The laminating cracks that are seen have not led to spalling and may therefore have been introduced by the diamond saw cutting the specimen.

The EDS provides the following chemical compositions at specified locations in weight percentages:

	Si	Ti	Al	V
Near the outer edge	47	50	0.8	1.7
Near the center	28	69	0.7	2.7
Ti alloy interface	1.5	92	3.7	2.8
Ti alloy farther in	1.5	93	3.6	2.4

The composition of the outer surface of the coating corresponds well to the phase $TiSi_2$, that of the inner zone of the coating to Ti_5Si_3. The diffusion zone at the interface with the substrate was too thin to make a determination but might be α-titanium because of the obvious depletion of aluminum in the coating. The compositions of the silicides do not exactly coincide with the published binary phase diagram, but that partly reflects the absolute accuracy of the instrument and interference from the alloying elements in the substrate metal. Etchant: 20% HF, 20% HNO_3, 60% glycerin.

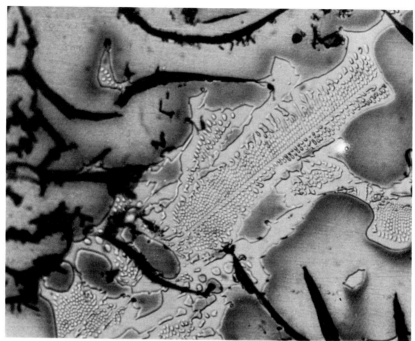

Fig. 7.5. Metal recovered from a secondary smelting operation is recognized as a ferronickel alloy with close to 30% nickel content. It has a melting point considerably less than the binary Fe–Ni phase diagram would indicate. The alloy has been found to be very hot-short. This might limit its use to a nickel source in the melting formulation of an alloy such as stainless steel. The origin of the hot-shortness might also indicate levels of impurities inimical to commercial alloys.

The micrograph (×1000) gives important evidence. A large amount of carbon has been absorbed, which explains the graphite flakes and is a partial reason for the low melting temperature. There is a well-formed eutectic, which is the most important reason for hot-shortness and a second reason for the low melting temperature. EDS results give the following compositions in weight percentages:

	Ni	Fe	Si	P
Eutectic	24.9	66.9	1.1	7.1
Dendrites	27.0	71.8	0.9	0.3

The proportions of the skeleton-shaped phase and matrix phase in the eutectic and the phosphorus analysis suggest an austenite–phosphide eutectic. In cast irons, this eutectic is called steadite. Phosphide eutectics in the Fe–P and Ni–P binary systems are 1050° and 880°C, respectively. This eutectic should be intermediate, so we have another source of the low melting temperature. The alloy will behave much like a cast iron. Etchant: 5 g $CuCl_3$, 10% HCl, 90% ethyl alcohol.

concentration changes often require additional methods of study. Hardness testing may be useful but only if the composition changes reflect substantial hardness differences and the depth of the surface changes is much greater than the dimensions of the hardness indentations.

In these situations EDS is particularly useful. Its microbeam dimensions permit point-by-point composition determinations so that a composition gradient can be drawn even if the surface zone is very thin. In many situations it is sufficient to identify the element or elements that have been enriched or depleted.

Composition gradients are implicit in the solidification process. The coring effect (see Chaper III) is a qualitative indication of a gradient of solute elements across a dendrite, but the dimensions and concentrations involved cannot be determined by ordinary metallography. Again, the ability to perform point-by-point analysis in succession across a specified distance is a valuable attribute of the EDS system. Diffusion associated with cladding manufacture (see Chapter V) or coatings exposed to service produces the same kind of condition and the condition can be assessed by EDS in the same way.

We present two examples of concentration gradients. One (Fig. 7.6) represents a surface effect and the other (Fig. 7.7) an assessment of the segregation of a solute element in the dendritic structure of a cast material.

Partition of Alloying Elements

A binary phase diagram allows an explicit definition of the compositions of both phases in a two-phase field, given the average composition of an alloy and the temperature. When third and fourth alloying elements are added the situation becomes unclear, not because phase diagrams cannot be constructed but because such constructions are rare. All too often, commercial alloys are polycomponent with both large and small solute contents. It is not necessary, most of the time, to have a substantial part of complex phase equilibria mapped out when the issues of solute distribution can often be explicitly defined by the use of EDS directed to each of the phases observable by metallography.

For example, phosphorus in a steel can exist in two forms. At a low concentration it can harden the ferrite phase considerably and, if in a tempered martensitic condition, strengthen the microstructure without jeopardizing the low ductile–brittle transition temperature. It can also contribute to the hardenability of a steel. At some concentrations, the austenite–iron phosphide eutectic exists with concomitant hot short-

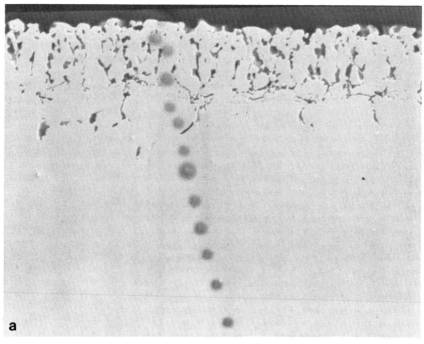

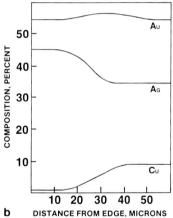

b   DISTANCE FROM EDGE, MICRONS

FIG. 7.6. The pre-Hispanic natives of the Central Andes in South America developed a method of enriching superficially the gold contents of copper–gold and copper–silver–gold alloys in fabricated form. This was probably accomplished by selective oxidation of the copper from the alloy by heating in air. (a) The scanning electron micrograph (×750) illustrates the surface after exposure of an alloy containing about 54% Au, 36% Ag, and 10% Cu to air at 700°C for an accumulated time of 40 h. The oxide crust was leached away by use of a dilute acid. The selective oxidation of the copper left a spongy surface that is almost copper-free. The SEM helps to bring out the appearance of the porosity. The accompanying graph (b) shows the results of point-by-point EDS determinations of the three alloy components. Because the sample was mounted in a nonconducting resin and analysis was performed at the sample edge, the polished surface was coated with carbon. The long exposure to the electron beam during analysis damaged the carbon film, leaving telltale marks.

254

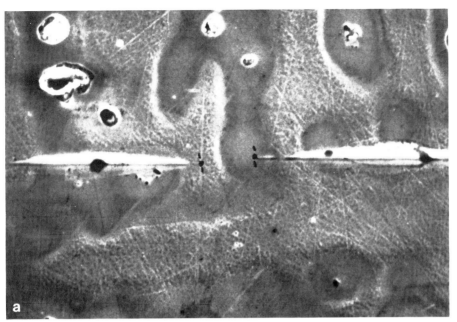

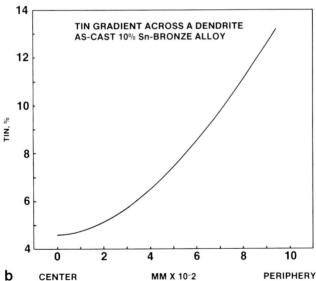

FIG. 7.7. (a) The coring effect in an as-cast, 10% tin-bronze is illustrated in the micrograph ($\times 1000$). The actual magnitudes of the tin gradient have been determined by marking out a distance between the center and the periphery of a dendrite using the ends of two microhardness indentations as shown. This distance is 0.1 mm. (b) Plot of the EDS point-by-point measurements. As might be expected from the phase diagram of the Cu–Sn system, the center of the dendrite is much leaner in tin than the average composition of the casting because the liquidus and solidus lines drop with increasing tin content from pure copper. The slope of that drop and the divergence of the liquidus and solidus govern the extent of segregation between the surface and center of a dendrite. Etchant: 5 g $FeCl_3$, 5% HCl, 95% alcohol.

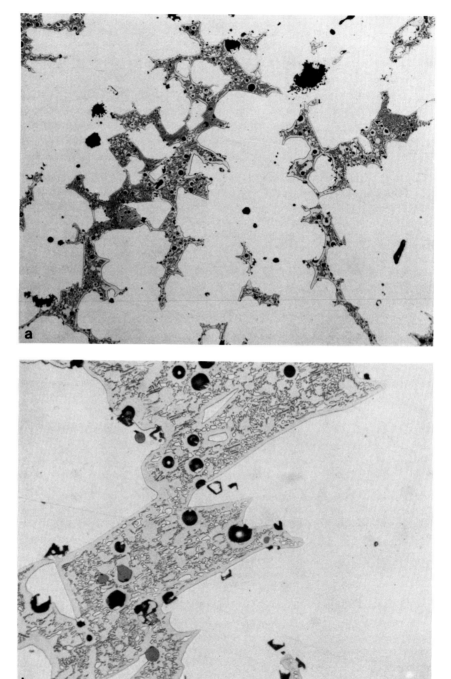

ness. Unfortunately, a small amount of eutectic is likely to adopt the divorced configuration (see Fig. 3.22) and now the problem is to distinguish a phosphide particle from a carbide in the microstructure. The microbeam of the EDS instrument makes this a fairly direct procedure. It is simply enough to see tall peaks of both phosphorus and iron in the element profile when the beam is centered on particles in question.

Diffusion processes in particular tend to generate segregations of solute elements that present unfamiliar metallographic configurations. The consequences of these local segregations cannot be considered without some knowledge of the element(s) engaged in the segregation process. As in the previous example, EDS indications of a qualitative nature are often sufficient.

We present two cases of alloy partitioning wherein light microscopy sets questions to be answered and other techniques including EDS are used to provide the information leading toward answers. Figure 7.8 represents the results of nonequilibrium solidification of a ternary alloy.

FIG. 7.8. (a) Microstructure ($\times 100$) of a leaded tin-bronze in the as-cast condition. The average composition is about 17% Sn and 1.5% Pb. The tin level is somewhat beyond the solubility limit of the copper solid solution according to the binary Cu–Sn equilibrium diagram. (b) The interdendritic microstructure ($\times 400$) is a jumble of perhaps three or four additional phases.

Lead has no known measurable solid solubility in copper, and the EDS verifies this. The interdendritic region appears to be eutectic but actually is eutectoid structure of the delta phase which lies as a band around the Cu–Sn solid–solution dendrites and as the matrix of the eutectoid structure. By EDS, the delta phase contains about 32% Sn. The delta matrix isolates a dispersion of a copper solid solution with lower Sn content. The black globules are lead that has acquired some tin (1.4%) and some copper (9.6%).

The purpose of the exercise was to find an explanation for the fact that alloying of lead with bronzes reduces their brittleness at high tin contents. The theory has been that the lead somehow reduces the percentage by volume of the brittle, continuous delta phase. The lead seems not to have done this at all, so the influence of lead on the amelioration of brittleness in high-tin bronze castings needs another, perhaps simpler, explanation. Etchant: 5 g FeCl$_3$, 10% HCl, 90% ethyl alcohol. The following EDS results were obtained for the three principal phases in a leaded tin-bronze (in weight percent):

	Dendrites	Rim and matrix	Distributed phase	Globules
Copper	88.5	67.7	84.95	9.6
Tin	10.4	31.3	14.0	1.4
Lead	0.1	0.1	0.1	88.1

The question of how the tin segregates in a ternary bronze is perhaps academic but no less challenging. Figure 7.9 represents a serious industrial problem that requires more information than ordinary good metallography can provide.

FIG. 7.9. A very large twist drill has been fabricated from a high-speed steel alloy for the cutting and fluted portion and a 0.45% carbon steel for the shank where it is gripped by the chuck of the drill press. The composite is joined by a flash butt weld. Dye penetrant techniques often show cracks in the weld zone, but even without such indications the weld can be fractured by banging the unfinished drill against an iron anvil.

Metallographically, the weld seems well and uniformly achieved as seen in the micrograph (a) ($\times 400$). In an arbitrary section it would be unexpected good fortune to find a crack and its root. One looks for a structure that is likely to be abnormally brittle. Close scrutiny shows the as-cast condition of the tool steel which has not been completely expelled as liquid during the pressure cycle of the welding operation. Another feature of the weld zone is better seen by the SEM view (b) ($\times 500$) of the etched structure. In most cases, light optical microscopy provides better contrasts and nuances than SEM with the same etched condition. However, polish relief, if it occurs, is better seen by SEM. From the scanning electron micrograph, the molten high-speed steel, which has a significantly lower melting temperature than the plain carbon steel, is seen to have penetrated into a grain boundary of the unalloyed carbon steel. The boundary must have been between austenite grains at the welding temperature.

The diluted forms of the tool steel are less problematic than the solidified structure of the diffusion-modified, highly alloyed tool steel. Tool steel manufacturers make every effort to supplant the as-cast ingot condition with a well-forged and rolled product for machining and heat treatment. Only the destruction of the coarse carbide network of the as-cast condition and a fine-grained, tempered martensitic structure after finishing the heat treatment can provide the necessary toughness for cutting-tool service. The as-cast structure rejuvenated by the welding operation has inserted a low-toughness condition precisely at the weld. Etchant: 2% HNO_3, 98% ethyl alcohol. EDS analysis of the alloy contents of the tool steel near the weld gave the following results (alloy compositions in weight):

	Remote[a]	Interface	Grain boundary penetration
Chromium	3.6	2.6	1.1
Molybdenum	8.6	6.0	4.1
Tungsten	2.8	1.3	1.7
Cobalt	6.8	4.2	1.5

[a] Seems close to the specification for the M30 grade.

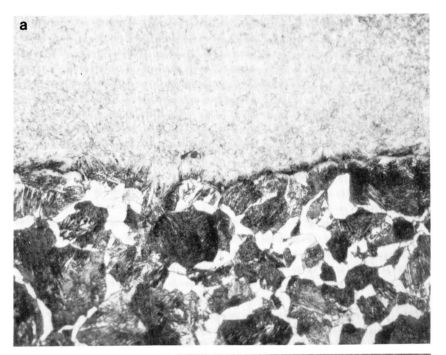

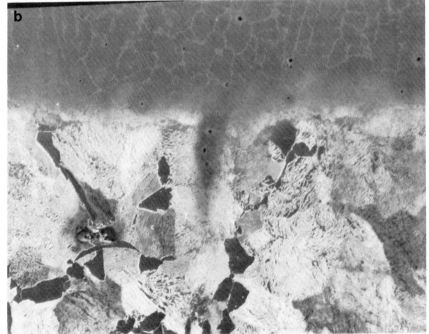

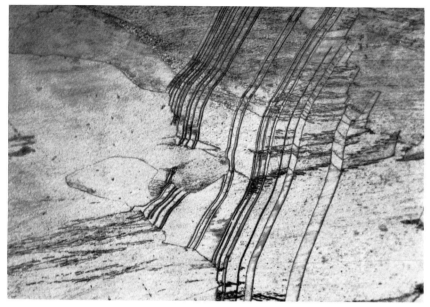

FIG. 7.10. Copper metal of geologic origin in the earth's crust is known as native copper. In many parts of the world, large deposits of native copper still remain. The metal is quite pure and deformable. The Indians of North America fabricated ornaments and tools from native copper long before the advent of the Europeans. The micrograph (×95) is of a specimen of native copper that had been thinned by about 58% by cold rolling. We see bands running across a very large grain of copper. The boundary etching intensity is very great, and it might seem that these bands represent distorted annealing twins. However, copper rarely forms more than two annealing twins per grain.

We consider these to be deformation bands. As discussed in the caption of Fig. 1.8, the orientation differences across deformation bands are insufficient to generate enhanced etching activity and so to be observable. Accordingly, we think that segregation of minor impurities to the interfaces of deformation bands might have had this effect. The elements in dilute solid solution were identified as iron, silver, and zinc in this specimen of native copper.

The accompanying table summarizes the EDS data (compositions in weight percent) for these elements at three positions: A, matrix well away from the deformation band boundaries; B, matrix between deformation band boundaries; C, on the deformation boundaries. The x-ray energies were enhanced by increasing the dwell time of the electron beam on the points. Several points were analyzed for each position class.

The evidence is that zinc segregated to the deformation band boundaries; to a lesser extent, so did iron. Such segregation is governed by the exposure to an elevated temperature, which in the present case was the temperature of the mold heating in the mounting press (ca. 150°C). Etchant: 33% NH_4OH, 33% of 3% H_2O_2, 34% water. (*Figure continues*)

Segregation at Interfaces

Metallographic examination sometimes reveals unusual or unexpected etching effects at grain boundaries and in symmetric lines across the interiors of large grains. What is evident is either an unusually rapid and deep etch at grain boundaries or the definition by etching of lines suggesting boundaries. Since etching and relative etching activity are often related to unusual composition differences, one wonders what they mean.

It is easy to assume that an exceedingly thin band of precipitates exists at the grain boundary. Precipitates usually have compositions very different from those of the grains on either side of the boundary, hence the very active etching. Such precipitates can be revealed by the magnifications provided by electron microscopy, but even at the limits of resolution of electron microscopy precipitate particles may not be seen.

It is increasingly evident that with dilute solutions of alloying elements, solute elements can segregate to grain interfaces without causing transformation of the crystal structure of the solvent metal. An example is the resistance to grain coarsening by silver added to copper in amounts of only ounces per ton.

There is also evidence that other forms of systematic lattice disregistry involve local regions of increased vacancy population that attract atoms of solute elements, resulting in concentration differences at and away from the sites of high vacancy population. Moreover, these segregations are considerable when the average concentrations are very small.

EDS can be used to provide evidence for such segregation and the identities of the segregating atoms. We present two cases (Fig. 7.10 and 7.11) of solute segregation where the metallographic evidence is ambiguous.

Fig. 7.10 (*Continued*)

| | Location | | |
	A	B	C
Iron			
Range	0.02/0.15	0.09/0.23	0.20/0.37
Average	0.08	0.18	0.25
Silver			
Range	0.00/0.24	0.00/0.10	0.00/0.16
Average	0.14	0.04	0.05
Zinc			
Range	0.10/0.63	0.00/0.49	0.57/1.01
Average	0.17	0.21	0.88

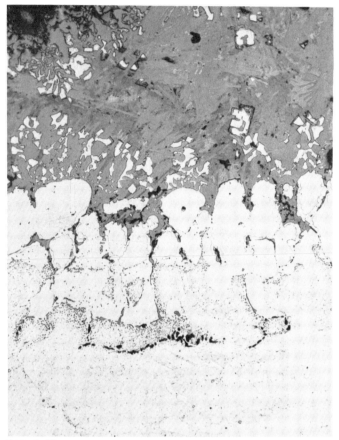

FIG. 7.11. Sulfur-containing environments, gaseous or liquid, are extremely destructive to nickel alloys. The most frequent event is embrittlement at elevated
temperatures. Hot-shortness of nickel alloys with exceedingly small sulfur contents
has been a common problem.

We immersed a sample of Inconel 600 alloy in a molten sulfide which is best
characterized as the mineral chalcopyrite. The nominal composition of Inconel 600 is
76% Ni, 15.5% Cr, and 8% Fe. The exposure of the Inconel 600 sample, a piece cut
from rolled plate, was at 900°C for about 24 h. The micrograph (×100) of the interface
between the nickel alloy and the solidified sulfide shows the exceedingly aggressive
attack. The molten sulfide can be seen in various stages of attack: intergranular
penetration, envelopment of grains, and dissolution of the grains.

To demonstrate that this reaction is not just superficial, we took an EDS beam to a
grain boundary far enough from the reaction zone that it is not clear whether any
sulfur penetration has occurred. Based on an average of 10 spot tests, the analysis is as
follows: 63.9% Ni, 13.8% Cr, 9.0% Fe, 2.7% Cu, and 8.2% S$ Indications are that the
penetrant along grain boundaries is a liquid rich in sulfur and the copper and iron
with which it was initially associated. Etchant: 5 g FeCl₃, 10% HCl, 90% ethyl
alcohol.

Nonmetallic Inclusions

Chapters I and VI discuss the role of nonmetallic inclusions in the mechanical properties of the metal in which they are embedded. Another part of the story deals with the understanding that nonmetallic inclusions are related to the stages of manufacture. In the liquid state, molten slags can be occluded in the metal while it is being poured and as the result of turbulence when the metal is in a semisolid state. During solidification, other kinds of nonmetallic inclusions develop as the result of interaction between solutes in the metal solution, for instance, deoxidation reactions.

Whichever is the source of nonmetallic inclusions, their compositions provide important clues to how the manufacturing processes were managed. In preindustrial times, the inability to prevent impurity buildup meant that a broad span of quality grades had to be sorted laboriously because the process was largely uncontrolled. Neutralization of impurities by controlled interaction with deliberate additions that drew the impurities into an insoluble condition made the nonmetallic inclusion the lesser of alternative evils. Modern approaches are based on the idea that limiting impurity absorption in the first place is a third option that is more cost effective in the long run.

Perhaps the analysis of nonmetallic inclusions is most rewarding in the reconstruction of ancient processes. In these materials the only residues of information are locked up in the identities and amounts of impurities in the metal and the nonmetallic inclusions embedded in and preserved by the metal.

In Fig. 7.12 we present a comparison of two wrought irons made by different processes, which dominated the production of general-purpose, formable iron at different times in European and American history. The "fingerprints" of the two processes are locked up in the minor components of the slag inclusions in the metal. The heavy-element analyses must be converted to oxide proportions to allow interpretation. The oxide identities of most metals as they exist in a slag are well known and singular. Thus, we expect Fe as FeO, Mn as MnO, Si as SiO_2, Al as Al_2O_3, Ca as CaO, P as P_2O_5, K as K_2O. However, iron can exist in both divalent and trivalent forms in slags and this cannot be recognized from the EDS analyses. In general, we must assume a divalent form of iron, but the assumption might make for a small correction in estimating slag fusion temperature. We present the metal analyses converted to oxide forms in the accompanying table.

Figure 7.13 represents a truly ancient copper from Cyprus. The information in sulfide inclusions bears on the mineral origins of the metal.

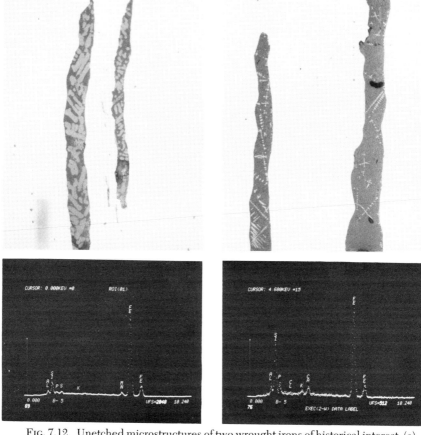

FIG. 7.12.   Unetched microstructures of two wrought irons of historical interest. (a) Micrograph (×248) of a sample of American origin; (b) micrograph (×124) of a sample recovered during excavation on a property on the coast of Jamaica. The microstructures of both samples are essentially the same and represent primary crystals of wüstite (FeO) in a matrix of fayalite (Fe_2SiO_4). The obvious difference is in the relative proportions of wüstite. The EDS analyses confirm this in the relative proportions of Fe and Si converted to their respective oxide forms but also show other differences, as may be seen in the EDS traces and in the table below. The most important difference is the absence of potassium (as potash) in one sample and its significant presence in the other.

This evidence distinguishes the processes used. Before the late 18th century, wrought iron was made by either reduction of iron ore by charcoal (bloom smelting) or decarburization of pig iron by drip melting it through a bed of charcoal ignited with an excess of air (fining). In both cases the metal was exposed to an occluded slag which dissolved the charcoal ash, containing, as does all wood, a significant amount of potash. (*Figure continues*)

This is of great interest to archeologists and historians because human history underwent great changes with the ability to extract copper from ores by smelting.

Corrosion Conditions

Since corrosion involves the interaction of a solid material with components of its environment, the characterization of each is necessary for a full explanation. The corroded specimen may be all that remains to diagnose the nature of the problem. We have two materials to examine: (1) the corrosion product if it is still adherent to the surface of the specimen and (2) the specimen itself and, in particular, the mode of attack—general or local. Metallography is the starting point. It can reveal even small residues adherent to the surface of the specimen and allow scrutiny by the EDS beam. Either light or electron microscopy can be used for this reconnoiter. It can also show special circumstances such as intergranular corrosion attack or pitting attack.

Fig. 7.12 (*Continued*).   During the 19th century, the bloom smelting and fining processes were gradually supplanted entirely by the puddling process. In this process a molten, shallow pool of pig iron was decarburized by oxygen-rich burned gases or iron oxides but out of direct contact with solid fuel. Thus puddled iron never had the opportunity to absorb potash via the occluded slag.

The compositions of slag inclusions in the two historical wrought irons are as follows:

American: early 20th century		Jamaican: 18th or 19th century	
EDA analysis	Conversion	EDA analysis	Conversion
Fe 91.0%	FeO 88.8%	Fe 78.0%	FeO 68.5%
Mn 2.8	MnO 2.7	Mn 0.5	MnO 0.4
Si 3.6	SiO_2 5.8	Si 11.6	SiO_2 17.5
Al 1.0	Al_2O_3 1.4	Al 2.0	Al_2O_3 2.7
Ca 0.1	CaO 0.1	Ca 2.6	CaO 2.6
P 0.5	P_2O_5 0.9	P 4.4	P_2O_5 7.1
K 0.0	K_2O 0.0	K 1.4	K_2O 1.2

Calculation procedure: $\%Fe \times 1.29 + \%Mn \times 1.29 + \%Si \times 2.14 + \%Al \times 1.89 + \ldots = S$, whence $\%Fe \times 1.29/S = \%FeO$ and so on.

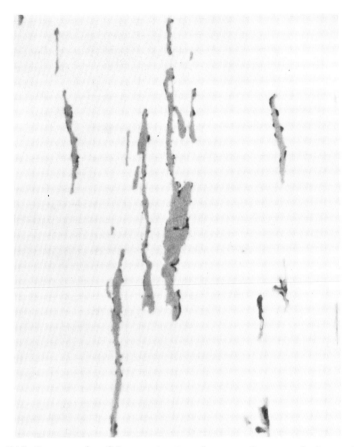

FIG. 7.13. Some results of the examination of a piece of copper which is perhaps 4000 years old, loaned by Tamara Stech (University of Pennsylvania). The non-metallic inclusions shown in the unetched microstructure ($\times 1000$) are sulfides containing predominantly copper and some iron (12.3% S, 86% Cu, 1.3% Fe). Like iron silicate slag inclusions in wrought iron, these sulfide inclusions are deformable to ribbon shapes at the hot working temperatures. Indeed, with this volume of sulfide inclusions, deformation processing was difficult to conduct at normal temperatures without destructive cracking.

The existence of sulfide inclusions, particularly in the amount apparent, signifies that the metal was recovered by smelting ores that were largely sulfide in nature. The smelting process involved the selective oxidation of iron from the fused mineral, which was most likely chalcopyrite. In the critical stage of smelting the molten sulfides (called matte) are heavily depleted of their iron (removed in slag) and self-reduce to copper metal by the action of oxygen from air or oxides.

The inclusions clearly show a minority phase, which is black compared to the gray majority phase. We infer from the ternary phase diagram for the system Cu−Fe−S that three phases coexist at low temperatures: copper metal itself, chalcocite (Cu_2S), and bornite, which has significant solubility for iron at low temperatures. That iron is present in the sulfide and not as elemental iron metal inclusions is a clear indication that the ore was not predominantly of the oxide type (as are malachite ores).

The EDS identifies the heavy elements but ignores the light elements (generally sodium and below). This leaves the question of what has been ignored. Most common corrosion events involve oxygenated water and the corrosion products are hydrated oxides of the elements in the metal specimen. The base metal element will dominate in the EDS spectrum, but the alloying elements may be less obvious because of the dilution with oxygen. Thus, a corroded stainless steel may show weaker indications of nickel and chromium than the alloy itself.

Aqueous corrosion can involve conversion to metal ions which, by secondary reactions, reorganize into new solids. In the corrosion of bronzes, the elements tin and lead, as well as copper, are converted to ions. But the tin reassociates to form oxides, while the copper and lead form carbonates or sulfates. Thus the corrosion can involve primary and secondary processes of reaction with the environment. We can use EDS evidence for elements such as chlorine or sulfur to point to special features of the corrosion environment such as the presence of NaCl or H_2S, which have significant solubilities in condensed water. Instead of basic carbonates, these intrinsic components of the environment lead to corrosion products that can be identified by x-ray diffraction as basic sulfates or basic chlorides. But the examination may not have to be carried that far if the chlorine or sulfur can be clearly identified in the corrosion products. The exact form of the corrosion products with these active agents may be generally known.

If the corrosion products are fugitive because they are fluids— gaseous or liquid—it is unlikely that the understanding of the corrosion reaction can be more than hypothetical. However, if the corrosion attack is intercrystalline or in the form of deep pits, as demonstrated by metallographic examination, vestiges of the corrosion products may remain sequestered in the cracklike roots of penetration. Also, such specialized corrosion patterns are often narrowly produced by very few environmental factors.

In some circumstances, only stains are residual from the corrosion event. More surface-sensitive methods of analysis, such as electron spectroscopy for chemical analysis (ESCA), serve in these cases. Since there is always a considerable element of conjecture about the nature of a corrosion event, metallography and EDS may often be more effective in ruling out certain mechanisms than in permitting more positive deductions.

Figure 7.14 represents an example of a heavy occluded element that was useful in defining the corrosion environment. Figure 7.15 is a case where the evidence of corrosive attack was clear but the agent was missing and could only be reasonably inferred.

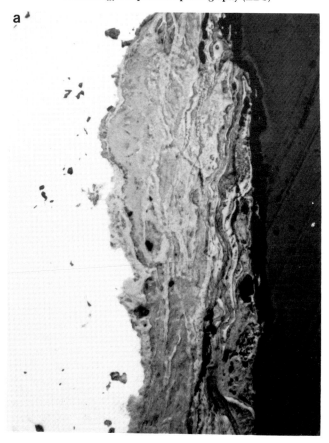

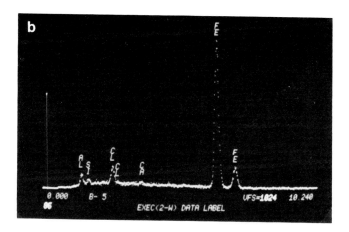

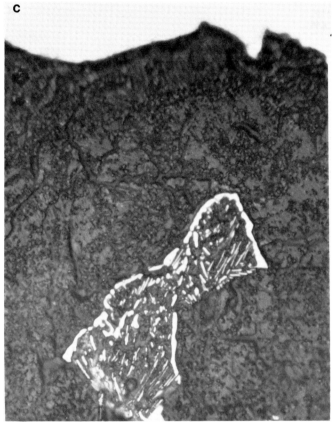

FIG. 7.14. (a) Micrograph (×50) showing an unusually dense rust deposit on the surface of a wrought iron object recovered during ground excavation near the coast of the island of Jamaica. (b) The EDS profile for the rust shows iron to be the dominant heavy element. The oxygen and hydroxyl entities in the rust are not disclosed, although their association with the iron atoms can be reasonably inferred. Chlorine makes a strong showing. Of course, chlorine cannot exist by itself; it must be bound with other light atoms, which are not detected. Because of the location of the discovery so near the sea, we can guess that sodium chloride is embedded in rust. Minor showings of calcium (probably as calcium carbonate) and silicon (probably as silica) are likely to represent extraneous minerals that infiltrated the rust (iron oxide–hydroxide) through ground waters. These minor elements may have provided the densifying factors to the rust body.

(c) This micrograph (×1000) is an unusual opportunity to see the corrosion process at an intermediate stage. We see an erstwhile colony of pearlite from the steel. The ferrite matrix in the pearlite colony has entirely corroded before the iron carbide (cementite) skeleton has begun to corrode. We think that the calcium, as calcium carbonate, filled the normally porous rust and stopped continuing corrosion. (*Figure continues*)

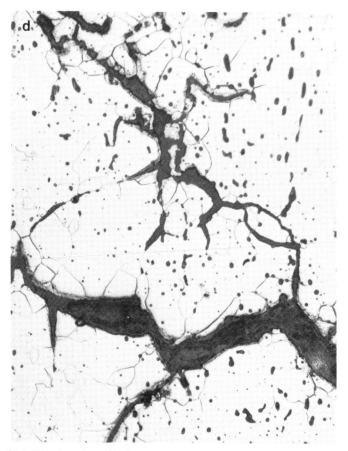

Fig. 7.14 (*Continued*).   (d) This micrograph (×200) shows another unusual situation. We see an intercrystalline pattern of corrosion in the metal. The yawning openings of the major pattern suggest stress corrosion cracking, but this is not a common attack on unalloyed iron. Whether by stress corrosion cracking or by intergranular corrosion, this microstructure rarely develops. The known environmental conditions for laboratory simulation are boiling sodium hydroxide or alkali nitrates. However, something that occurs in days in laboratory testing should be possible in years at ordinary temperatures. Sodium hydroxide is an unlikely chemical, but nitrates are a common derivative of barnyard animal discards.

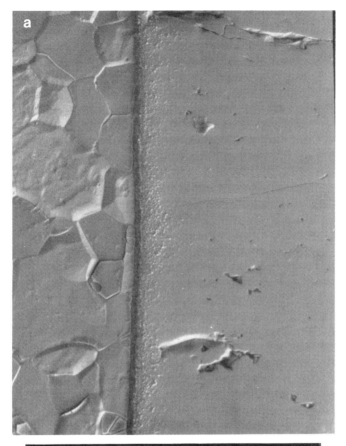

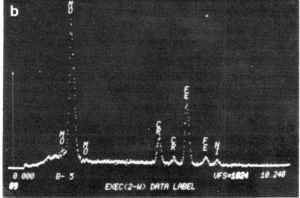

FIG. 7.15.   (*Figure continues*)

FIG. 7.15. A molybdenum heating element wound in a flat accordion fashion from rod of about 5-mm diameter is one of several suspended from the walls and roof of a furnace used to braze parts assemblies made from stainless steel. The heating element was seen to thin at the bottom of the loops in the hottest zones of the furnace. Some solidified droplets of a metal were evident. The presumption was that the heating element was being attacked by a vapor that contained metal, the metal alloying with the molybdenum when the vapor was reduced to these elements. This improbable alloying considerably lowered the melting temperature of the molybdenum, which drained to the lowest points. The low-melting liquid metal continued to dissolve molybdenum at the lowest point. As the molybdenum element thinned, its electrical resistance increased and so did the local temperature. The process was self-accelerating.

The scanning electron micrograph (a) ($\times 300$) showed a sharp boundary between the simple polycrystalline structure of molybdenum and the alien deposit. The EDS spectrum (b) shows that the components of the alloy, in addition to Mo, were Fe, Cr, and Ni, which are the metal components of the stainless steel parts being brazed out of contact with the heating elements. The measured proportions of these elements were also close to those of the stainless steel.

The brazing temperature was not so high that appreciable vapor pressures of iron, chromium, and nickel existed. These elements had to be transported as volatile salts. A study of the manufacturing steps showed that parts were machined, degreased, and assembled on trays for brazing. Both the cutting fluids used in machining and the degreaser were chlorinated hydrocarbons. If inadequately degreased, threads and other remote crevice-like arrangements could sequester chlorinated, grease-like entities. At elevated temperatures the stainless steel was attacked to form metal chlorides, which are quite volatile. The furnace atmosphere contained a high proportion of hydrogen gas for the purposes of (1) reducing oxide coatings on the parts (fluxing) so that the brazing action can proceed rapidly and (2) preventing oxidation of the heating elements.

Hydrogen can also reduce metal chlorides to metal, but the thermodynamics are only marginally favorable. However, if the reduced metals can diffuse (dissolve) into molybdenum, the reactions between hydrogen gas and metal chloride vapors can proceed much more rapidly even if the thermodynamics are unfavorable.

The phase equilibria for the Mo–Fe–Cr alloy system show that certain zones of composition possess melting temperatures comparable to those in the furnace at the surfaces of the heating elements. This accounts for the formation of liquid metal droplets and the gradual thinning of the heating wire. The remedy for the problem was to purify the trichlorethylene degreaser bath more frequently.

For general reading on subjects in this chapter see references (8) and (9). Other topics covered in the references are scanning electron microscopy (1,2,6,7), energy dispersive spectrography (3,5–7,12), sample preparation (4–6), and quantitative analysis (4–6,8–11).

REFERENCES

1. G. Van der Voort, in "Applied Metallography."pp. 139–170. Van Nostrand Reinhold, New York, 1986.
2. J. Verhoeven, Scanning electron microscopy, in "Metals Handbook" (9th ed.) vol. 10, pp. 490–515. American Society for Metals, Metals Park, Ohio, 1986.
3. K. Heinrich and D. Newberry, Electron probe x-ray microanalysis in "Metals Handbook" (9th ed.) vol. 10, pp. 515–535. American Society for Metals, Metals Park, Ohio, 1986.
4. H. Yakowitz, Methods of quantitative x-ray analysis, in "Practical Scanning Electron Microscopy," pp. 327–372. Plenum Press, New York, 1975.
5. H. Yakowitz and J. Goldstein, Practical aspects of x-ray microanalysis, in "Practical Scanning Electron Microscopy," pp. 401–434. Plenum Press, New York, 1975.
6. L. Murr, Electron and ion probe microanalysis, in "Electron and Ion Microscopy and Microanalysis," pp. 141–165. Marcel Dekker, New York, 1982.
7. E. Lifshin, X-ray generation and detection in the SEM, in "Scanning Electron Microscopy," pp. 243–276. McGraw-Hill, New York, 1974.
8. K. F. J. Heinrich, "Electron Beam X-Ray Microanalysis", Van Nostrand Reinhold, New York, 1981.
9. J. I. Goldstein, D. E. Newbury, P. Echlin, D. C. Joy, C. Fiori, and E. Lifshin, "Scanning Electron Microscopy and X-Ray Microanalysis", Plenum Press, New York, 1981.
10. C. E. Fiori and D. E. Newberry, *Scanning Electron Microscopy*,1,401–422 (1978).
11. R. B. Marinenko, K. F. J. Heinrich, and F. C. Ruegg, Microhomogeneity studies of NBS standard reference materials, "NBS Research Materials, and Other Related Samples," NBS Special Publication No. 260–65, National Bureau of Standards, Gaithersburg, MD (1979).
12. D. E. Newberry, and S. Greenwald,*J. Res., Natl. Bur. Stand.***85**,429–440 (1980).

Supplementary Reading
Recommendations

An attempt has been made here to provide the reader with additional sources of information and technical discussion on the general subject of metallography. The list of references is certainly not complete, but many of these are themselves reference sources. The intention of this Appendix, therefore, is to provide a point of initiation for some detailed search of the literature or for further reading.

Topic	*References*
Principles of microscope optics	1, 2, 3, 4, 5, 6, 7, 69, 70
Microscructures	57, 58, 59, 63, 64, 65, 66, 72, 73, 82, 93, 94
Related periodicals.	67, 68
Etching and etchants	2, 7, 9, 10, 12, 13, 14, 15, 69, 70, 72, 75, 76, 84, 98
Etch pit techniques	14, 15, 16, 17, 18, 19, 20
Electropolishing and chemical polishing	13, 14, 69, 70, 72, 75, 76, 80, 81, 87, 91, 92
Optical microscopy	57, 58, 59, 62, 64, 65, 69, 70, 72
Sample preparation	2, 5, 7, 8, 9, 10, 11, 12, 13, 57, 58, 59, 63, 69, 70, 71, 72, 75, 76, 77, 78, 79, 81, 83, 84, 85, 86, 88, 89, 90
Scanning electron microscopy	57, 58, 59, 60, 61, 62, 64, 66
Grain size standards and methods of determination	
austenitic	2, 9, 21, 22
ferritic	21
nonferrous	2, 9, 23
Iron and steel	
macroexamination	2, 24, 25
inclusion ratings and identification	9, 21, 26, 27, 28, 95, 96, 97
microstructure of transformations	29, 30, 31
etchants	7, 9, 10, 22
Cast iron	
graphite flake size and arrangements, standards	21, 32

Topic	References
eutectic cell size	33, 34
microstructures	31, 32, 35, 36
Techniques for special materials, metals and alloys	
aluminum and its alloys	7, 10, 22, 31, 36, 37, 38
antimony	10
bearing metals	10
beryllium	7, 10, 39
cemented carbides	10
ceramics	40
chromium	10, 41
cobalt	7, 10, 28
columbian (niobium)	42, 43
copper	7, 9, 10, 22, 31, 36, 44
germanium	7
hafnium	45
indium	10
lead	9, 10, 22, 31
magnesium	9, 10, 22, 31, 36, 46, 47
precious metals	7, 9, 10, 22
refractory metals	7, 10, 56
silicon	10
tantalum	10, 42, 43
thorium	7, 48
tin	7, 9, 10, 22
titanium	10, 49, 50
uranium	10
vanadium	7, 51
zinc	7, 9, 10, 22, 36
zirconium	10, 52, 53, 54, 55

REFERENCES

1. A. B. Winterbottom, in "The Physical Examination of Metals" (B. Chalmers and A. G. Quarrell, eds.), Chapter I. Arnold, London, 1960.
2. G. L. Kehl, "Principles of Metallographic Laboratory Practice." McGraw-Hill, New York, 1949.
3. G. L. Kehl, in "Modern Research Techniques in Physical Metallurgy," Chapter 1. ASM, Metals Park, Ohio, 1953.
4. S. Tolansky, "Surface Microtopography." Wiley (Interscience), New York, 1960.
5. R. M. Allen, "Photomicrography" (2nd ed.). Van Nostrand, Princeton, New Jersey, 1958.
6. "Photography through the Microscope," Kodak Publ. No. P-2. Rochester, New York.
7. "Metals Handbook," Supplement: Electrolytic polishing and etching, pp. 169–177. ASM, Metals Park, Ohio, July 15, 1954.
8. H. S. Cannon, A universal polishing method. *Metal Progr.* **67**, 83–86 (1955).
9. *Am. Soc. Testing Mater., ASTM Std.* (Part 1B), pp. 803–859 (1946).

10. C. Smithells, "Metals Reference Book" (3rd ed.) Vol. 1, pp. 214–266. Butterworths, London, 1962.
11. W. J. McG. Tegart, "The Electrolytic and Chemical Polishing of Metals in Research and Industry." Pergamon, Amsterdam, 1956.
12. C. A. Johnson, Electrolytic etching, Metals Dig. 1, Nos. 2, 3 (1955); Aspects of electrolytic polishing, Metals Dig. 1, Nos. 5, 55 (1955); Electrolytic polishing for metallography, Metals Dig. 4, Nos. 1, 4 (1958); Effect of pressure in mechanical polishing, Metals Dig. 4, Nos. 1, 4 (1958); Effect of pressure in mechanical polishing. Metals Dig. 4, Nos. 3, 2 (1958).
13. P. A. Jacquet, Electrolytic and chemical polishing, Met. Rev. 1, 157–238 (1956).
14. C. E. Morris, Electropolishing of steel in chrome–acetic acid electrolyte. Metal Progr. 56, 696–699, 712, 714 (1949) (etch pit procedures).
15. J. W. Mitchell, in "Direct Observations of Imperfections in Crystals" (J. B. Newkirk and J. H. Wernick, eds.). Chapter 1. Wiley (Interscience), New York, 1962.
16. H. G. F. Wilsdorf, Observations of dislocations. Natl. Bur. Std. Monogr. 59 (1963).
17. J. R. Low and R. W. Guard, The dislocation structure of slip bands in iron, Acta Met. 7, 171–179 (1959).
18. R. W. Guard, An etch pit method for revealing dislocation sites in nickel, Trans. AIME 218, 573–574 (1960).
19. J. J. Gilman, Etch pits and dislocations in zinc monocrystals, Trans. AIME 206, 998–1044 (1956).
20. J. D. Meakin and A. G. F. Wilsdorf, Dislocations in deformed single crystals of alpha brass, Trans. AIME 218, 737–745 (1960).
21. Am. Soc. Testing Mater. ASTM Std. (Part I, Ferrous metals) (1952).
22. "Metals Handbook." ASM, Metals Park, Ohio, 1948.
23. F. C. Hull, A new method for making rapid and accurate estimates of grain size, Trans. AIME 172, 439 (1947).
24. "Metals Handbook," Supplement, pp. 195–200. ASM, Metals Park, Ohio, 1955.
25. G. A. Roberts, J. C. Hamaker, and A. R. Johnson, "Tool Steels," pp. 70–84. ASM, Metals Park, Ohio, 1962.
26. M. Baeyertz, "Non-Metallic Inclusions in Steel." ASM, Metals Park, Ohio, 1947.
27. S. L. Case and K. R. VanHorn, "Aluminum in Iron and Steel," pp. 52–77. Wiley, New York, 1953.
28. C. H. Lund and H. J. Wagner, Identification of Microconstituents in Superalloys, Defense Materials Information Center Memorandum No. 160, Nov. 15, 1962 (available from office of Technical Services, U.S. Department of Commerce, Washington, D.C.).
29. E. C. Bain and H. W. Paxton. "Alloying Elements in Steel." ASM, Metals Park, Ohio, 1961.
30. J. H. G. Monypenny, "Stainless Iron and Steel," Vol. 2. Chapman & Hall, London, 1954.
31. R. M. Brick and A. Phillips, "Structure and Properties of Alloys." McGraw-Hill, New York, 1949.
32. H. T. Angus. "Physical and Engineering Properties of Cast Iron," Brit. Cast Iron Res. Assoc., Birmingham, England, 1960.

33. J. V. Dawson and W. Oldfield, Eutectic cell count—An index of metal quality, *J. BCIRA* **8**, 221–231 (1960).
34. H. D. Merchant, Metallography of eutectic cells in cast iron, *Foundry* **91**, 59–65 (1963).
35. A. Boyles, "Structure of Cast Iron." ASM, Metals Park, Ohio, 1947.
36. G. Lambert, "Typical Microstructures of Cast Metals," Brit. Inst. Foundrymen Publ., Manchester, England, 1957.
37. L. F. Mondolfo. "Metallography of Aluminum Alloys." Wiley, New York, 1943.
38. F. Keller and G. W. Wilcox, Polishing and etching of constituents of aluminum alloys. *Metal Progr.* **23**, 45 (1933); see also "Metals Handbook," pp. 798–803. ASM, Metals Park, Ohio, 1947.
39. M. C. Udy, *in* "The Metal Beryllium." ASM, Metals Park, Ohio, 1955.
40. W. D. Kingery, "An Introduction to Ceramics," pp. 402–458. Wiley, New York, 1960.
41. W. D. Forgeng and G. T. Motoch, *in* "Ductile Chromium." ASM, Metals Park, Ohio, 1957.
42. G. L. Miller, "Tantalum and Niobium." Academic Press, New York, 1959.
43. W. D. Forgeng. *in* "Columbium and Tantalum" (F. T. Sisco and E. Epremian, eds.), pp. 507–534. Wiley, New York, 1963.
44. F. H. Wilson, *in* "Copper, The Metal, Its Alloys and Compounds" (A. Butts, ed.), pp. 873–884. Reinhold, New York, 1954.
45. F. M. Cain, Jr. *in* "Zirconium and Zirconium Alloys." ASM, Metals Park, Ohio, 1953.
46. A. Beck, "Technology of Magnesium and Its Alloys." F. A. Hughes, London, 1940.
47. "Magesium Laboratory Methods." Dow Chemical Co. Publ., Midland, Michigan, 1957.
48. H. P. Roth, *in* "The Metal Thorium." ASM, Metals Park, Ohio, 1958.
49. A. D. McQuillan and M. K. McQuillan, "Titanium." Butterworths, London, 1956.
50. H. R. Ogden and F. C. Holden, Metallography of Titanium Alloys, Titanium Metallurgical Laboratory Report No. 103, May 29, 1958 (available from office of Technical Services, U.S. Dept. of Commerce, Washington, D.C.).
51. W. Rostoker, "The Metallurgy of Vanadium." Wiley, New York, 1958.
52. H. P. Roth, Metallography of zirconium, *Metal Progr.* **58**, 709–711 (1950).
53. A. H. Roberson, Metallography of zirconium and zirconium alloys, *Metal Progr.* **56**, 667–669 (1949).
54. P. A. Jaquet, Electrolytic polishing of zirconium, *Metallurgia* **42**, 268–270 (1950).
55. M. L. Picklesimer, Anodizing as a Metallographic Technique for Zirconium-Base Alloys, ORNL-2296. TID4500 (13th ed.) (available from office of Technical Services, U.S. Dept. of Commerce, Washington, D.C.).
56. F. R. Cortes, Electrolytic polishing of refractory metals, *Metal Progr.* **80**, 97–100 (1961).
57. R. Gray and J. McCall, Microstructural Science, Vol. 1, *Proc. 5th Ann. Mtg. IMS.* Elsevier, New York, 1974.
58. G. Fritzke, J. Richardson, J. McCall, Microstructural Science, Vol. 2. *Proc. 6th Ann. Mtg. IMS.* Elsevier, New York, 1974.

59. P. French, R. Gray, J. McCall, Microstructural Science, Vol. 3, *Proc. 7th Ann. Mtg. IMS.* Elsevier, New York, 1975.
60. O. C. Wells, Scanning Electron Microscopy. McGraw-Hill, New York. 1974.
61. V. Colangelo and F. Heiser, Analysis of Metallurgical Failures. Wiley, New York, 1974.
62. J. McCall and W. Mueller, Metallographic Specimen Preparation—Optical and Electron Microscopy, *Proc. IMS–ASM Symp. 1973.* Plenum, New York, 1974.
63. J. McCall and W. Mueller, Microstructural Analysis—Tools and Techniques, *Proc. IMS–ASM Symp. 1972.* Plenum, New York, 1973.
64. "Metals Handbook," Vol. 7, Atlas of microstructures of industrial alloys. ASM, Metals Park, Ohio, 1972.
65. "Metals Handbook," Vol. 8, Metallography, structures and phase diagrams. ASM, Metals Park, Ohio, 1973.
66. "Metals Handbook," Vol. 9, Fractography and atlas of fractographs. ASM Metals Park, Ohio, 1974.
67. Praktische Metallographie. Verlag and Anzeigenannahme, Stuttgart, Germany.
68. *Metallography,* Journal of IMS. Elsevier, New York.
69. R. C. Gifkins. "Optical Microscopy of Metals." Elsevier, New York, 1974.
70. J. H. Richardson, "Optical Microscopy for the Material Sciences." Dekker, New York, 1971.
71. L. E. Samuels, "Metallographic Polishing by Mechanical Methods." Elsevier, New York, 1971.
72. L. Habraken and J. L. DeBrouwer, DeFerri Metallographia, Vol. 1: Fundamentals of metallography. Saunders, Philadelphia, 1966.
73. A. Schrader and A. Rose, De Ferri Metallographia, Vol. 11: Structures of steel. Saunders, Philadelphia, 1966.
74. A. Pokorny and J. Pokorny, DeFerri Metallographia. Vol. III: Solidification and deformation of steels. Saunders, Philadelphia, 1966.
75. ASTM Standards Part II, Metallography: Nondestrictuve Testing. ASTM, Philadelphia, 1975.
76. ASTM STP557 Mettallography. ASTM, Philadelphia, 1974.
77. A. Suzirmae and R. M. Fisher, Specimen Damage During Cutting and Grinding, pp. 3–9. Techniques of Electron Microscopy, Diffraction and Microprobe Analysis, ASTM STP372 (1963).
78. L. E. Samuels, An improved method for the taper sectioning of metallographic specimens, *Metallurgia,* **51**, 161–162 (Mar. 1955).
79. A. E. Calabra. Mounting metallographic specimens for edge retention, *Metal Progr.,* **90**, 103–104, 110 (July 1966).
80. R. Pinner, Theory and practice of chemical polishing: A survey of solutions and processes for various applications, *Electroplating Metal Finishing* **7**, 127–131, 140 (1954).
81. J. M. Dickinson, Polishing hard metals electromechanically, *Metal Progr.* **74**, 142–144 (Oct. 1958).
82. C. J. Smithells, Metals Reference Book, (4th ed.) Vol. 1. Butterworths, London (1967).
83. J. T. Salmon, *The Microscope* **8**, (6), 139 (1951).
84. R. H. Greaves and H. Wrighton, Practical Microscopical Metallography (4th ed.). Chapman Hall, London (1967).

85. A. E. Calabra. *Metals Progr.* **90**, 103 (1966).
86. E. W. Filer and J. P. Smith, *Metals Progr.* **83**, (5), 126 (1963).
87. G. Gutzeit, Electroplating Engineering Handbook (2nd ed.) p. 471. Reinhold, New York (1962).
88. W. C. Coons, *Metal Progr.*, **49**, 956 (1946).
89. E. M. Nelson, *J. Roy. Microscop. Soc.* **5**, 713 (1885).
90. J. H. Richardson, *Metallography*, **1**, 149 (1968).
91. F. C. Hull and R. L. Anderson, "Tentative Methods for Electrolytic Polishing of Metallographic Specimens." Westinghouse Research Laboratories, Scientific Paper 6-99402-11PZ (April 1958).
92. P. A. Jacquet, Electrochemical techniques in modern metallography, *Proc. Am. Electroplaters Soc.*, 240–254 (1959).
93. J. Nutting and R. G. Baker. "The Microstructure of Metals." Brit. Inst. Metals Publ., 1965.
94. L. Dillinger *et al.*, Microstructure of Heat Resistant Alloys. Alloy Castings Inst. Publ., 1970.
95. "Sulfide Inclusions in Steel," Materials/Metalworking Technology Series No. 6. ASM, Metals Park, Ohio, 1975.
96. T. R. Allmand, Microscopic Identification in Steel. Brit. Iron and Steel Assn. Publ., 1962.
97. R. Kiessling and N. Lang, "Non-Metallic Inclusions in Steel," Brit. Inst. Iron and Steel, Spec. Rep. No. 90, 1964.

Index